Harwood Fundamentals of Pure and Applied Economics

# GAME THEORY IN INTERNATIONAL ECONOMICS

# FUNDAMENTALS OF PURE AND APPLIED ECONOMICS



---

## INTERNATIONAL ECONOMICS I
In 3 Volumes

| | | |
|---|---|---|
| I | Game Theory in International Economics | *Mcmillan* |
| II | Uncertainty and the Theory of International Trade | *Grinols* |
| III | Disequilibrium Trade Theories | *Itoh and Negishi* |

# GAME THEORY IN INTERNATIONAL ECONOMICS

JOHN McMILLAN

ROUTLEDGE
Taylor & Francis Group

First published in 1986 by
Harwood Academic Publishers GmbH

Reprinted in 2001 by
Routledge
2 Park Square, Milton Park, Abingdon, Oxon, OX14 4RN

Transferred to Digital Printing 2009

*Routledge is an imprint of the Taylor & Francis Group*



The publishers have made every effort to contact authors/copyright holders of the works reprinted in *Harwood Fundamentals of Pure & Applied Economics*. This has not been possible in every case, however, and we would welcome correspondence from those individuals/companies we have been unable to trace.

These reprints are taken from original copies of each book. in many cases the condition of these originals is not perfect. the publisher has gone to great lengths to ensure the quality of these reprints, but wishes to point out that certain characteristics of the original copies will, of necessity, be apparent in reprints thereof.

*British Library Cataloguing in Publication Data*
A CIP catalogue record for this book
is available from the British Library

Game Theory in International Economics
ISBN10: 0-415-26909-1 (hbk)
ISBN10: 0-415-48840-0 (pbk)

ISBN13: 978-0-415-26909-4 (hbk)
ISBN13: 978-0-415-48840-2 (pbk)

International Economics I: 3 Volumes
ISBN 0-415-26908-3
Harwood Fundamentals of Pure & Applied Economics
ISBN 0-415-26907-5

# Game Theory in International Economics

**John McMillan**

*University of California, San Diego, USA*

**A Volume in the International Trade Section**
**Edited by**
**M Kemp**
*University of New South Wales, Australia*

harwood academic publishers
Switzerland • Australia • Belgium •
France • Germany • Great Britain • India
Japan • Malaysia • Netherlands • Russia •
Singapore • USA

First published 1986
Second printing 1994

Harwood Academic Publishers

Private Bag 8
Camberwell, Victoria 3124
Australia

3-14-9, Okubo
Shinjuku-ku, Tokyo 169
Japan

12 Cour Saint-Eloi
75012 Paris
France

Emmaplein 5
1075 AW Amsterdam
Netherlands

Christburger Str. 11
10405 Berlin
Germany

820 Town Center Drive
Langhorne, Pennsylvania 19047
United States of America

Post Office Box 90
Reading, Berkshire RG1 8JL
Great Britain

---

**Library of Congress Cataloging in Publication Data**
McMillan, John, 1951–
Game theory in international economics.
(Fundamentals of pure and applied economics,
ISSN 0883-2366; vol. 1. International trade section)
Bibliography: p.
Includes index.
1. International economic relations—Mathematical models. 2. Game theory. I. Title. II. Series:
Fundamentals of pure and applied economics; v. 1.
III. Series: Fundamentals of pure and applied economics.
International trade section.
HF1412.M38 1986 337'.01'5193 86-3141
ISBN 3-7186-0277-6

---

ISBN 3-7186-0277-6. ISSN: 0883-2366. 

# CONTENTS

# INTRODUCTION TO THE SERIES

Drawing on a personal network, an economist can still relatively easily stay well informed in the narrow field in which he works, but to keep up with the development of economics as a whole is a much more formidable challenge. Economists are confronted with difficulties associated with the rapid development of their discipline. There is a risk of "balkanisation" in economics, which may not be favourable to its development.

*Fundamentals of Pure and Applied Economics* has been created to meet this problem. The discipline of economics has been subdivided into sections (listed on the inside back cover). These sections include short books, each surveying the state of the art in a given area.

Each book starts with the basic elements and goes as far as the most advanced results. Each should be useful to professors needing material for lectures, to graduate students looking for a global view of a particular subject, to professional economists wishing to keep up with the development of their science, and to researchers seeking convenient information on questions that incidentally appear in their work.

Each book is thus a presentation of the state of the art in a particular field rather than a step by step analysis of the development of the literature. Each is a high level presentation but accessible to anyone with a solid background in economics, whether engaged in business, government, international organizations, teaching, or research in related fields.

Three aspects of *Fundamentals of Pure and Applied Economics* should be emphasized:

—First, the project covers the whole field of economics, not only theoretical or mathematical economics.

—Second, the project is open-ended and the number of books is not predetermined. If new interesting areas appear, they will generate additional books.
—Last, all the books making up each section will later be grouped to constitute one or several volumes of an Encyclopaedia of Economics.

The editors of the sections are outstanding economists who have selected as authors for the series some of the finest specialists in the world.

# Game theory in international economics

JOHN McMILLAN[1]
*(Revised July, 1984)*

## 1. STRATEGY AND TRADE

THE THEORY OF GAMES provides a set of mathematical techniques for analyzing situations in which each agent's utility depends not only on his own actions but also on the actions of others; and all of the agents take these interdependencies into account when deciding their actions. Von Neumann and Morgenstern explained the difference between a game-theoretic problem and an optimization problem of the sort more traditionally solved in economic theory, such as a single consumer's maximization problem:

> Consider now a participant in a social exchange economy. His problem has, of course, many elements in common with a maximum problem. But it also contains some, very essential, elements of an entirely different nature. He too tries to obtain an optimum result. But in order to achieve this, he must enter into relations of exchange with others. If two or more persons exchange goods with each other, then the result for each one will depend in general not merely upon his own actions but on those of the others as well. Thus each participant attempts to maximize a function (his above-mentioned "result") of which he does not control all variables. This is certainly no maximum problem, but a peculiar and disconcerting mixture of several conflicting maximum problems. Every participant is guided by another principle and neither determines all variables which affect his interest. This kind of problem is nowhere dealt with in classical mathematics [159, pp. 10–11].

Game theory has had little influence on the theory of international economics: little mention of game theory is to be found in the trade-theory textbooks. Despite game theory's lack of impact on the body of trade theory, many of the important policy issues of

1. I owe thanks to Murray Kemp for initiating this project and for his encouragement and advice; to Avinash Dixit for helpful comments; to Harold Winter for his enthusiastic research assistance; and to the Social Sciences and Humanities Research Council of Canada for research support.

international economics have some game-theoretic character: negotiations over mutual reductions of tariffs, either bilaterally or under the General Agreement on Tariffs and Trade; the international indebtedness and threatened default of some less-developed countries, such as Brazil and Mexico; the formation and preservation of customs unions, such as the European Economic Community; issues of international common property, addressed in the Law of the Sea Conference; the establishment of cartels to raise the price of internationally traded commodities, such as the Organization of Petroleum Exporting Countries; the extent to which competition from foreign firms removes the need to use antitrust provisions to restrain domestic monopolies; the international implications of domestic macroeconomic policies, such as high interest rates in the United States or persistent Japanese balance-of-payments surpluses; the international coordination of policies to stabilize the volatile prices of the exports of less-developed countries; the sharing of costs and responsibilities within international alliances, such as the North Atlantic Treaty Organization; the possible international redistributions of income considered in the North-South debate; the use of trade as a weapon in political warfare, such as the sanctions imposed on Rhodesia and the Soviet Union; international problems of pollution, for example the pollution of the Rhine affecting each of the countries it flows through, or acid rain falling on both the United States and Canada. In all of these examples, there is strategic interdependence: what one agent's best action is depends upon what another agent does, and vice versa.

A common assumption in the theory of international trade is that countries are small: their exports and imports form such small fractions of total world trade that they cannot affect world prices. For many of the commodities imported or exported by most countries, this assumption is realistic. In some cases, however, a country's trade in some particular commodity does comprise a significant proportion of world trade, so that the small-country assumption is inapplicable. As soon as the small-country assumption is dropped, questions which are inherently game-theoretic arise.

This monograph surveys several applications of game theory in international economics. It does not exhaust the list of questions which have been, or could be, analyzed using game theory. The topics covered were chosen both to illustrate the range of questions in international economics to which game theory can be fruitfully applied, and to exemplify various concepts of game theory. The monograph should be useful both to the specialist in international economics seeking to learn some game theory, and to the game theorist looking for a new and very fertile area in which to apply the techniques of game theory.

The analysis is presented as nontechnically as possible. This means that the results are not necessarily expressed in their full generality, or with all the necessary technical caveats. Nor are all of the results that can be extracted from these models presented: instead, an attempt has been made to convey the flavor of the analysis. Although no sophisticated mathematics are used in this monograph, some subsections, marked with asterisks (*), present material which is either somewhat more technical than the other subsections or not essential to understanding the concepts to be introduced. These subsections prove results stated in preceding subsections; on a quick reading, these subsections can be omitted without loss of continuity (though with some inevitable loss of precision).

Section 2 presents those basic concepts of game theory which seem to be the most useful to economists: the noncooperative equilibrium and some cooperative-game solution concepts. Section 3 looks at a more specialized question in game theory: in a dynamic game, cooperative behavior may occur in a noncooperative context. The remaining sections describe some particular applications of game theory in international economics. Section 4 models as a game the interactions among different nations' tariff policies. Section 5 considers possible government reactions to the oligopoly power of foreign and domestic firms. Section 6 reviews the evidence on international cartels as a case study of game-playing in practice. Section 7 considers the possibility of decentralized stabilization of world commodity prices. Section 8 models the exploitation of international common property. Section 9 examines the formation of customs unions. Section 10 summarizes some models of international transfers. Section 11 identifies an equivalence between a basic simplifying assumption used in trade theory (the community preference function) and a basic simplifying assumption used in game theory (transferable utility). Finally, Section 12 discusses some further questions of international economics to which game theory is applicable.

Game-theoretic concepts used include the static Nash equilibrium (in Section 4 and Section 10), the Stackelberg equilibrium (in Section 5), the repeated game (in Section 6), the stochastic game (in Section 7), the differential game and perfect equilibrium (in Section 8), the core (in Section 9), and transferable utility (in Section 11).

### 1.1 Bibliographic notes

For general surveys of the theory of international trade, see Chipman [26], Kemp [75], Dixit and Norman [37], and Woodland [164].

## 2. BASIC CONCEPTS OF GAME THEORY

This section introduces some of the more important ideas of game theory. No attempt is made at completeness; only those ideas which seem to be the most relevant to economics are discussed.

### 2.1 Definitions

The distinctive feature of a game is the presence of interdependencies among the agents: one agent's utility depends not only on his own actions, but also on the actions of each of the other agents. It is the agents' awareness of interactions among their decisions which give rise to the subtle problems of game theory. (To take an example from elementary-textbook economics, the decisions of a monopolist are not game theoretic; the decisions of an oligopolist are game theoretic.)

A *strategy* is a complete description of the agent's planned actions. (In the case of the textbook oligopolist, a strategy is simply a level of output; and the set of strategies consists of all non-negative output levels.)

Formally, a *game* (or, more accurately, a *game in strategic form*) consists of: a set of agents; a set of strategies for each agent, from which the agent chooses a particular strategy to play; and a utility function for each agent. Suppose there are n agents. Each agent, $i$, chooses a strategy $a_i$ from the set $A_i$ of possible strategies. Because it is a game, each agent's utility, $u^i$, depends on every agents' strategy: $u^i = u^i(a_1, \ldots, a_n)$, $i = 1, \ldots, n$.

An agent uses a *mixed strategy* if he deliberately randomizes his choice of strategy. If, on the other hand, the agent decides nonstochastically, he is said to use a *pure strategy*. Mixed strategies are used when agents have some incentive to conceal what they are going to do.

In a *zero-sum game*, the sum of all agents' utilities is always zero. A zero-sum game is a game of pure conflict: what one agent wins, some other agents must lose. Clearly a *constant-sum game* is equivalent, with appropriate normalization, to a zero-sum game. A game of distributing a pie of fixed size is a zero-sum game; in most games in economics, however, the size of the pie depends upon how it is divided.

*Nonzero-sum games* have elements of both conflict and cooperation. It is in the agents' mutual interest to reach an outcome where the sum of utilities is relatively high; but the agents' interests conflict over their shares in that sum. (The textbook oligopoly problem is an example of a nonzero-sum game. The firms have a common interest in keeping total output low and therefore market price high. But it is in each firm's individual interest to have a large market share.)

Depending on the institutional context, a nonzero-sum game may be either cooperative or noncooperative. A game is *cooperative* if agents can, before acting, costlessly make agreements which are binding and enforceable. It is *noncooperative* if there are no institutions which ensure that the agents will keep any promises they make. (In nineteenth century Germany, collusive contracts made between firms were enforceable by the legal system: the oligopoly game was a cooperative game. In the present-day United States, such agreements could not be enforced; indeed, they are illegal: the oligopoly game is a noncooperative game.)

An allocation of resources is *Pareto optimal* (or *efficient*, or on the *utility-possibility frontier*) if there is no alternative allocation in which some agents are better off and none are worse off. It is important to note that here Pareto efficiency is defined with respect to the agents in the particular game; it does not imply efficiency in any wider sense. (Thus a Pareto optimal outcome in the oligopoly game occurs if the industry has somehow achieved a collusive outcome and total industry profit is maximized. This is clearly not efficient from the point of view of society as a whole.)

## 2.2 Cooperative games

There is a presumption that the equilibrium of a cooperative game will be Pareto optimal. The argument is simple. If the outcome is not Pareto optimal, then, by definition, there is some other outcome which all agents prefer (or at least some prefer it and the rest are indifferent). From a point inside the utility-possibility frontier like point A in Figure 2.1, there is a Pareto-improving movement such as in the direction of the arrow. If binding agreements can costlessly be made, Pareto-improving

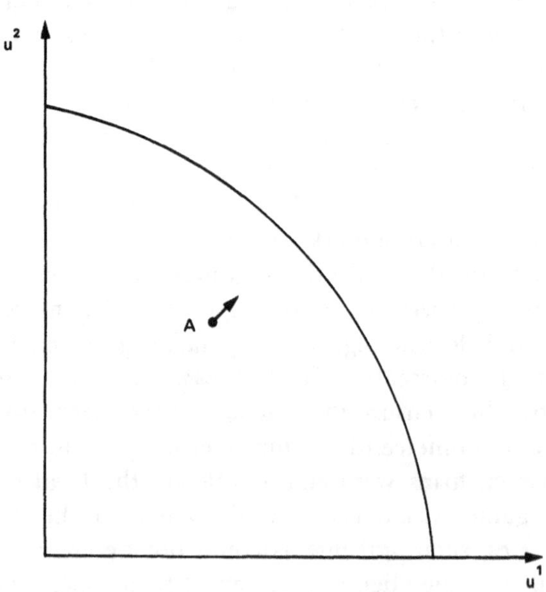


FIGURE 2.1 A Pareto-improving change.

changes will presumably occur; a non-Paretian point could not be an equilibrium of a cooperative game. Of course, a game can have many Pareto optimal points: theories of cooperative games attempt to select particular Paretian outcomes. (In the oligopoly example, if the legal system enforces cartel agreements, the outcome will be on the profit-possibility frontier. There remains, however, the bargaining problem of how the market is to be divided among the firms.)

The game theory literature contains many alternative solution concepts for cooperative games, essentially *ad hoc* ways of selecting specific Pareto optimal points. Two of these, the core and the Shapley value, will be briefly mentioned here.

A *coalition* is a collection of some or all of the agents in the game. Let $\mathcal{N}$ represent the set of agents; then a coalition $S$ is a subset of $\mathcal{N}$. Assume, for this discussion of cooperative games, that utility is *transferable*; that is, the utilities of the different agents can be added together, so that the utility of a coalition can be defined as the sum of the utilities of the agents in the coalition. (The restrictions on the form of the utility functions under which this is possible will be discussed in Section 11.5.) Then the *characteristic function* is a numerical measure of the maximal utility a coalition can attain for itself if there is no cooperation with agents outside the coalition. Note that a considerable amount of the structure of the game—the transactions that occur within the coalition—is hidden within the characteristic function. Denote the characteristic function by $v$; thus $v(S)$, for $S \subseteq \mathcal{N}$, is a number.

The *core* is the set of outcomes such that no coalition is able to make all its members better off. More precisely, the vector of utilities $(u_1, \ldots, u_n)$ is in the core if it is feasible and

$$\sum_{i \in S} u^i \geqslant v(S), \text{ for all } S \subseteq \mathcal{N}. \tag{2.1}$$

Note any allocation in the core must be Pareto optimal because one of the possible coalitions is $\mathcal{N}$, the set of all players.

The core, while it might be empty, often is very large. Consider now an alternative solution concept, which generates a unique solution. Let the marginal contribution of an agent $i$ to a coalition $S$ mean the difference between the value of the coalition with agent $i$ and without agent $i$; that is, $v(S) - v(S - \{i\})$. The *Shapley value* assigns to agent $i$ a level of utility equal to a weighted average of his marginal contributions to all possible coalitions. Specifically, the Shapley value is given by the formula

$$u^i = \sum_{S \subseteq N} \frac{(s-1)!(n-s)!}{n!} [v(S) - v(S - \{i\})] \tag{2.2}$$

(where $s$ is the number of elements of $S$). The weights in this weighted average can be interpreted as probabilities that particular coalitions will form if agents meet each other at random.

### 2.3 Noncooperative equilibrium

Consider a noncooperative game played once only. All agents have perfect information about the actions available to them and to their rivals. Moreover, each agent knows every other agent's utility function.

A strong notion of noncooperative equilibrium is the *dominant equilibrium*. At a dominant equilibrium, each agent's best action is uniquely defined regardless of any other agent's action. An example of a game which has a dominant equilibrium is the prisoner's dilemma. Two agents each have two possible strategies, $C$ ("cooperate") and $D$ ("defect"). The utilities are shown in Table I, the first number in each pair representing the utility of agent 1. (The actual numbers are immaterial; all that matters is their ranking.) If agent 1 predicts his rival will play $C_2$, his best action is to play $D_1$; if he predicts his rival will play $D_2$, his best action is again $D_1$. Agent 2 reasons symmetrically. Regardless of what an agent thinks his rival will do, he rationally plays $D$; $D$ is the dominant strategy. Notice that at the equilibrium each agent

TABLE I
The Prisoners' Dilemma

| | $C_2$ | $D_2$ |
|---|---|---|
| $C_1$ | (10, 10) | (0, 11) |
| $D_1$ | (11, 0) | (1, 1) |

receives a utility of 1, even though there is an alternative outcome in which each receives a utility of 10.

Most games do not have a dominant equilibrium; a less stringent notion of equilibrium is required. At a *Nash equilibrium*, each agent does the best he can given the other agents' actions. That is, $a^N \equiv (a_1^N, \ldots, a_n^N)$ is a Nash equilibrium if, for all $i = 1, \ldots, n$,

$$u^i(a_1^N, \ldots, a_{i-1}^N, a_i^N, a_{i+1}^N, \ldots, a_n^N) \\ \geqslant u^i(a_i^N, \ldots, a_{i-1}^N, a_i, a_{i+1}^N, \ldots, a_n^N) \text{ for all } a_i \in A_i. \tag{2.3}$$

No agent can do better than play his Nash strategy given that all the other agents are playing their Nash strategies. Given appropriate differentiability assumptions, and assuming the $a_i$'s are single-dimensional so that $A_i$ is a subset of the real line, the Nash equilibrium $(a_1^N, \ldots, a_n^N)$ is found by solving n simultaneous equations:

$$\frac{\partial u^i}{\partial a_i} = 0, \text{ for all } i = 1, \ldots, n. \tag{2.4}$$

It is of interest to compare a Nash equilibrium with a Pareto optimum. At a Pareto optimum, no agent can be made better off without some other agent being made worse off: $\sum_{j=1}^n \omega_j u^j$ is maximized for some constants $\omega_1, \ldots, \omega_n$ (with $\omega_k \geqslant 0$ for all $k$, $\omega_k > 0$ for some $k$); that is, at an interior Pareto optimum,

$$\sum_{j=1}^n \omega_j \frac{\partial u^j}{\partial a_i} = 0, \text{ for all } i = 1, \ldots, n. \tag{2.5}$$

Denote a Pareto optimum by $a^* \equiv (a_1^*, a_2^*, \ldots, a_n^*)$.

Suppose that, at a Pareto optimum, the different agents' interests conflict, in the sense that if any agent were to increase his action $a_i$, this would increase his own utility but lower every other agent's utility: that is,

$$\left.\frac{\partial u^i}{\partial a_i}\right|_{a^*} \geqslant 0 \text{ and } \left.\frac{\partial u^j}{\partial a_i}\right|_{a^*} < 0 \quad \text{for all } j \neq i,\ i, j = 1, \ldots, n. \tag{2.6}$$

This is a sufficient condition for the utility-possibility frontier to be negatively sloped: a change which increased one agent's utility decreases all the other's utilities. With condition (2.6), it follows from equation (2.5) that $\partial u^i/\partial a_i > 0$ at a Pareto optimum. Thus the sets of equations (2.4) and (2.5) cannot be satisfied simultaneously: the Nash equilibrium is not Pareto optimal. In this noncooperative game, as in the prisoners' dilemma, there is an inconsistency between individual and collective rationality: it is in each agent's interest that a Pareto optimal outcome be attained; but each individual's pursuit of his self-interest drives the system to a Pareto inferior equilibrium. Suppose the system is at a Pareto optimal point. At this point, $\partial u^i/\partial a_i > 0$, $i = 1, \ldots, n$ (from (2.5) and (2.6)). Thus it is in each agent's interest to act differently, provided no other agent acts differently. But all reason in this way, and so none plays Pareto optimally. Suppose, on the other hand, that the system is at a Nash equilibrium. Then, since the Nash equilibrium is not Pareto efficient, there is some alternative set of actions which would make all agents better off. By the definition of Nash equilibrium, however, it is in no agent's interest unilaterally to act differently.

Despite this feature of Nash equilibria, the case for Nash equilibrium as a solution concept is compelling. If the system were not at a Nash equilibrium, some individual could do better by acting differently.

(Returning to the oligopoly example, interpret the agents as firms, the $a_i$'s as outputs, and the $u^i$'s as profit functions. The Nash equilibrium described by the equations (2.4) is then just the familiar Cournot equilibrium: at the Cournot-Nash equilibrium, each firm equates its own marginal revenue, given the outputs of the other firms, to its marginal cost. Points which are Pareto optimal with respect to the firms in the industry, that is, outcomes on the profit-possibility frontier, correspond to cartel outcomes. The result that Pareto optimal outcomes are not Cournot-Nash equilibria means that each firm has an incentive to cheat on any collusive arrangement. This is sometimes misleadingly called the cartel instability problem: misleadingly because it is a problem of equilibrium, not stability as the terms are usually used in economics. A demonstration of the inefficiency of Cournot-Nash equilibrium, alternative to that provided by (2.4) through (2.6), proceeds as follows. Consider Cournot's oligopoly model: $n$ firms produce a homogeneous commodity, $a_i$ is firm $i$'s output, the industry demand function is $f(\sum_{j=1}^{n} a_j)$, and the common constant average cost is $c$. The $i^{\text{th}}$ firm's marginal revenue is $a_i f' + f$, and the industry's marginal revenue is $(\sum_{j=1}^{n} a_j) f' + f$. Since $f'$, the slope of the demand curve, is negative, the firm's marginal revenue exceeds the industry's. If a cartel maximizes total industry profit, it equates industry marginal revenue to the common marginal cost. But then any firm's marginal revenue exceeds its marginal cost; it is in the individual firm's interest to produce more than its share of the cartel output. The firm, unlike the cartel, does not take into account the lower profits of the other firms caused by the lower price resulting from its own higher output.)

### 2.4 Statics, dynamics, and timing

The foregoing description of Nash equilibrium assumed the agents made their actions simultaneously (or, what is essentially the same, the agents committed themselves to particular actions before observing the others' actions). A different equilibrium, the *Stackelberg equilibrium*, is obtained if one agent, the *Stackelberg leader*, can commit himself to an action before the other agents. The others, the *Stackelberg followers*, after observing the leader's action, choose their best actions. The Stackelberg equilibrium is solved for in two stages (the solution process going in the opposite direction to the actual sequence of events). First is computed the optimal responses of the followers to any action of the leader. This produces a set of equations the same as (2.4), except that there are only $n-1$ equations, one for each of the followers. Second, using these responses of the followers, the leader's optimal action is found. The leader picks the best outcome for him from the set of outcomes satisfying the $n-1$ equations for the followers' first-order conditions.

A common misunderstanding, propagated by intermediate microeconomics textbooks, of the meaning of Cournot-Nash equilibrium is based on a confusion between statics and dynamics.

A dynamic adjustment process is sometimes appended to the Cournot oligopoly model. Firms are presumed to conjecture some fixed level of output for each of their rivals. Then they choose their best replies to these conjectured outputs, each firm solving equation (2.4). Away from equilibrium, the outputs the firms conjecture for their rivals will be different from what the rivals actually produce. Conjectures are revised and the process is repeated until, if the system is stable, it reaches a Cournot-Nash equilibrium, at which conjectures are confirmed. This adjustment process is often correctly criticized because the firms' behavior is strangely myopic: at each time-point each firm conjectures that its rivals' outputs are fixed, despite repeatedly during the disequilibrium process observing such conjectures to be falsified.

It is incorrect to deduce from this that Cournot-Nash behavior is necessarily myopic or naive. The Cournot-Nash equilibrium need not be based on such a dynamic process. The model of Section 2.3 is static. Nowhere in the model does time enter: actions take place only once. If the Nash equilibrium is unique, then, given the assumption that each agent knows his rivals' strategy sets and utility functions, there is no need for an agent to be ignorant of his rivals' planned action. Each agent can imagine himself in the position of any other agent and can deduce what the other agent will do. The mutual consistency nature of the Cournot-Nash equilibrium (each agent is doing the best he can given that the other agents are playing their Cournot-Nash strategies) means that each agent rationally plays his Cournot-Nash strategy. The expectation that others will play Cournot-Nash is a rational expectation.

This rational-expectations justification for playing Nash strategies breaks down if, as is possible, the Nash equilibrium is not unique. If there is more than one Nash equilibrium, then it is not enough to determine an agent's best action for the agent to know his rivals' strategy sets and utility functions and to know that his rivals are rational: the agent will not be able to predict which strategies his rivals will choose. Game theory offers no systematic way of choosing a particular equilibrium when there are multiple equilibria. All that can be done is, in an *ad hoc* way, to add new elements to the theory in order to predict which equilibrium will be reached.

The criticism of the dynamics in the stability analysis of Cournot equilibrium was that the agents were myopic: they ignored the fact that they were in a dynamic process. Modelling a game in explicitly dynamic terms requires that each agent maximizes the present value of his utility stream (and not just current utility as in equation (2.4)). In this case, as will be discussed in Section 3.2, there may be other equilibria in addition to the static Cournot-Nash equilibrium.

## 2.5 Cooperative versus noncooperative games in economics

Should a particular economic situation be modelled as a cooperative game or as a noncooperative game? The answer depends on the institutional context: on whether it can be ensured that any agreements made are enforced; on whether, that is, there is a law of contract backed by some enforcement mechanism. For example, the so-called Coase Theorem says, in part, that when an externality from one agent's actions affects another agent, rational agents will negotiate so as to reach a Pareto optimal solution. One interpretation of this result is that it is simply an application of the basic result of cooperative game theory: an equilibrium of a cooperative game is Pareto optimal. On this interpretation, it is clear that the Coase Theorem holds only in certain institutional settings. Questions of international economics, for example, are probably best modelled as noncooperative games: there is little international law, and no international police force to enforce contracts. Equilibria of noncooperative games must now be further investigated.

## 2.6 Bibliographic notes

The seminal references on game theory are von Neumann and Morgenstern's *Theory of Games and Economic Behaviour* [159] and Nash's papers on cooperative games [112] and noncooperative games [113], Cournot [28] and Edgeworth [42] were economists who anticipated some of the ideas of modern game theory (respectively, the Nash equilibrium and the core).

For an enjoyable, nontechnical introduction to the game-theoretic style of thinking, see Schelling [141]. As a general survey of game theory, Luce and Raiffa [96] remains indispensable. Modern treatments of game theory, with some economic applications, are given by Friedman [44] and Shubik [146]. On the mathematics of game theory, see Owen [125].

Sufficient conditions for existence and uniqueness of Nash equilibrium were given by Friedman [44], Chapter 7. That static Nash equilibria are usually inefficient was shown by Dubey [38].

On the rationality of playing Nash equilibrium strategies in a static context, see Johansen [68]. Friedman [44] emphasized the need for care in distinguishing static games from dynamic games.

The interpretation of the Coase Theorem given in Section 2.5 is due to Arrow [5].

## 3. DYNAMIC GAMES

Sections 2.2 and 2.3 revealed an important difference between the outcome of a cooperative game and the outcome of a noncooperative game. If the institutional setting is such that a game is cooperative, then the agents will fully exploit the gains from trade and reach a Pareto optimal equilibrium. If, however, the game is noncooperative, each agent's pursuit of his self interest results in an equilibrium which is usually Pareto inferior. The games in Sections 2.2 and 2.3 were static. This section shows that the distinction between cooperative-game equilibria and noncooperative-game equilibria is blurred in dynamic games: in a dynamic noncooperative game, there may be equilibria which are Pareto optimal. For this to happen a specific type of dynamics must be introduced: the agents adopt strategies in which their current actions depend upon the past actions of the other agents. Strategies of this form allow the possibility of retaliation, and so cooperative behavior can be enforced in a noncooperative context. This insight must now be given precise content.

### 3.1 The Folk Theorem

Consider a game of the sort modelled in Section 2.3, but now imagine that same game to be played, not just once, but repeatedly, once in each time-period over an infinite time-horizon. (It is crucial for the solution to be described that neither the utility functions $u_i$ nor the sets of possible actions $A_i$ change over time, and that utility be a function of current actions alone, not past actions.) Each agent can observe the past actions of the other agents, and can use this information about the history of the game in making his current decision. In this dynamic context, a strategy for agent i is a plan of actions for each time period, $a_{i1}$, $a_{i2}, \ldots, a_{it}, \ldots$ (the second subscripts denoting time), where the action at time $t$, $a_{it}$, is a function of all agents' actions in some or all previous time periods $(a_{1\tau}, a_{2\tau}, \ldots, a_{n\tau})$ for $\tau \leqslant t-1$. This dynamic game is called a *repeated game* or a *supergame*.

Suppose each agent seeks to maximize some kind of average level of utility over time (that is, future utility is not discounted). Denote as *individually rational* any outcome which gives each agent at least as much utility as he can ensure for himself by his own actions. Then a result of obscure origin, known as the Folk Theorem, says that any individually rational outcome (and in particular, any Pareto optimal outcome) can be maintained as a Nash equilibrium in the repeated game.

This result is of limited interest to economists, because it requires future returns not to be discounted. However, a similar result holds in the more conventional case of discounting. Suppose agent $i$ seeks to maximize the present value of his utility stream, $\sum_{t=1}^{\infty} \alpha_i^{t-1} u^i(a_{1t}, a_{2t}, \ldots, a_{nt})$ (where $\alpha_i$, $0 \leqslant \alpha_i < 1$, is $i$'s discount rate). There can be many equilibria in this repeated game with discounting.

Note first that, if all the agents other than i play their static-equilibrium strategies $a_j^N, j \neq i$ (found by solving equations (2.4)), then clearly the best $i$ can do is also to play $a_i^N$. Thus the repeated playing of a static Nash equilibrium is a dynamic Nash equilibrium.

As with the Folk Theorem, there can be many other dynamic Nash equilibria; in contrast to the Folk Theorem, however, it is necessary to add the caveat that the discount rate must not be too high. For example, a Pareto optimum which all agents prefer to the static Nash equilibrium can be attained as an equilibrium of the repeated game if all agents adopt strategies of "threatening" to "punish" agents who deviate from the prescribed Pareto optimal action. More precisely, suppose each agent in the first period plays his Pareto optimal action. In subsequent periods, each agent continues to play Pareto optimally unless he sees some agent deviating. He then responds to this deviation in the next period and subsequent periods by himself reverting to his static noncooperative action. It is shown in the next section that, provided each agent values future utility sufficiently highly, such a threat is enough to dissuade any agent from deviating from his Pareto optimal action. The repeated playing of a Pareto optimum is a dynamic Nash equilibrium.

### 3.2 Dynamic equilibria

Denote by $(a_1^*, a_2^*, \ldots, a_n^*)$ a Pareto optimum which all agents prefer to the static Nash equilibrium $(a_1^N, a_2^N, \ldots, a_n^N)$. (Although the static Nash equilibrium need not be unique, assume in what follows that it is unique.) The dynamic strategy (or *reaction function*) just mentioned can be written as, for each agent $j$, $j = 1, \ldots, n$,

$$a_{j1} = a_j^*$$

$$a_{jt} = \begin{cases} a_j^* \text{ if } a_{i\tau} = a_i^*,\ \tau = 1, 2, \ldots, t-1,\ i \neq j,\ i = 1, \ldots, n \\ \text{otherwise.} \end{cases} \tag{3.1}$$

Such a strategy vector is an equilibrium if no agent ever wants to change his action from $a_i^*$. Consider agent $i$'s gains and losses from changing his action (let the current period be $t = 1$).

If agent $i$ continues forever to play $a_i^*$, and all the other agents follow the strategy (3.1), then agent $i$'s discounted utility is

$$\sum_{t=1}^{\infty} \alpha_i^{t-1} u^i(a_1^*, \ldots, a_n^*) = \frac{1}{1-\alpha_i} u_i(a_1^*, \ldots, a_n^*). \tag{3.2}$$

Instead, agent $i$ could play some other action in period 1. Since the other agents could not respond to agent $i$'s deviation until the next period, this deviation would yield some immediate gains to agent $i$. This is because, from conditions (2.5) and (2.6), $\partial u^i / \partial a_i \neq 0$ at a Pareto optimum. There will be some action, different from $a_i^*$, which maximizes agent $i$'s utility given that all the other agents are playing their Pareto optimal actions $a_j^*, j \neq i$. Denote this deviating action for agent $i$ by $a_i''$, $a_i''$ is found by solving

$$\frac{\partial u^i(a_1^*, \ldots, a_{i-1}^*, a_i, a_{i+1}^*, \ldots, a_n^*)}{\partial a_i} = 0 \tag{3.3}$$

By deviating in this way, $i$ can earn high utility now; he must trade this off against low utility in the future when the other agents retaliate by playing $a_j^N, j \neq i$. Thus the discounted utility of agent $i$ if he deviates in period 1 is

$$u^i(a_1^*, \ldots, a_i'', \ldots, a_n^*) + \sum_{t=2}^{\infty} \alpha_i^{t-1} u^i(a_1^N, \ldots, a_n^N)$$

$$= u^i(a_1^*, \ldots, a_i'', \ldots, a_n^*) + \frac{\alpha_i}{1-\alpha_i} u^i(a_1^N, \ldots, a_n^N) \tag{3.4}$$

Agent $i$ will not deviate from playing Pareto optimally if the former gain exceeds the latter gain; comparing (3.2) and (3.4), playing Pareto optimally is a dynamic equilibrium (or *reaction-function equilibrium*) if, for all $i = 1, \ldots, n$:

$$\frac{1-\alpha_i}{\alpha_i} \leqslant \frac{u^i(a_1^*, \ldots, a_i^*, \ldots, a_n^*) - u^i(a_1^N, \ldots, a_i^N, \ldots, a_n^N)}{u^i(a_1^*, \ldots, a_i'', \ldots, a_n^*) - u^i(a_1^*, \ldots, a_i^*, \ldots, a_n^*)} \tag{3.5}$$

Provided the future is not discounted too heavily (or provided the short-run gains from playing $a_i''$ are not too high relative to the returns from playing $a_i^*$), (3.5) will be satisfied and the Pareto optimal outcome $(a_1^*, \ldots, a_n^*)$ can be sustained as an equilibrium.

The denominator of the right-hand side of (3.5) shows the extra utility of agent i if he deviates rather than continuing to play Pareto optimally; it shows the short-run gain from deviating. The numerator shows the per-period gain to agent i of being at the Pareto optimum rather than the static Nash equilibrium; it shows the long-run opportunity cost of deviating. The right-hand side of (3.5) is an index of cooperation for agent $i$; the larger is this index of cooperation, the more likely (in some sense) it is that agent $i$ will use the strategy (3.1). If all agents do this, the Pareto optimal outcome is reached.

There is a surfeit of equilibria. There will in general be many Pareto optima, and so many equilibria supported by strategies of the form (3.1). Indeed, attention need not have been restricted to

Pareto optimal points; any outcome which all agents prefer to the static Nash equilibrium can be maintained in a reaction-function equilibrium (provided the equivalent to condition (3.5) holds). And, as already noted, the repeated playing of the static Nash equilibrium is itself an equilibrium of the repeated game.

Notice that the foregoing argument relies on the time horizon being infinite. Suppose instead that the game is repeated only a finite number of times. Suppose, moreover, the static Nash equilibrium is unique. Then at the start of the last period, each player knows there is nothing his rivals can do to retaliate to his current action (since there is no future period). But therefore each player knows there is no credible way of deterring deviation in the second-last period; and so on, back to the first period. Thus the strategies described above for attaining Pareto optimal outcomes as non-cooperative-game equilibria break down. It does not follow, however, that Pareto optimal outcomes cannot be equilibria of finitely repeated games. It can be shown that cooperative-like equilibria may be attainable in finite-horizon games if the static Nash equilibrium is not unique; or if the agents exhibit a particular kind of uncertainty about the strategies their rivals are following; or if the definition of equilibrium in the repeated game is weakened to admit approximate equilibria.

### 3.3 Perfect equilibrium

Suppose, in a two-agent version of the game of the previous section, one agent, say agent 1, has deviated and played $a_1''$. What incentive does the other agent, 2, have to carry out the strategy (3.1) and play his retaliatory static Nash action $a_2^N$ in the next period? If he believes that agent 1 may revert next period to playing Pareto optimally $a_1^*$, then agent 2 is only harming himself by carrying out his threat of retaliation.

Consider a stronger notion of equilibrium. *A subgame perfect equilibrium* has the property (in this context) that it is in every agent's interest to follow his equilibrium strategy whatever action has actually been played in previous periods by the other agents. (Contrast this with the Nash equilibrium, in which it is in every agent's interest to follow his equilibrium strategy provided the other agents in previous periods followed their equilibrium strategies.)

For the Pareto optimal equilibrium just discussed to be perfect, it must be in the interest of an agent actually to carry out his threat to retaliate to another agent's deviation. Perfectness rules out empty threats: if agent 1 is to be dissuaded from deviating, he must believe that agent 2 will in fact retaliate; it must be in agent 2's interest to retaliate. This is ensured, and the range of outcomes attained as Nash equilibria in the previous section are attained as perfect equilibria, if strategies involve the entire history of the game in the following way. There is an infinite string of threatened retaliations: if agent 2 does not retaliate to agent 1's deviation then agent 1 in turn retaliates for the lack of retaliation, and so on. Thus agent 2 is given an incentive to carry out his threatened retaliation.

### 3.4 Conjectural variations

As an alternative to modelling oligopoly as a static or dynamic game, use is sometimes made of the concept of *conjectural variations*. In a static framework, the oligopolists are modelled as conjecturing the existence of a relationship between their own output choices and those of their rivals. The conjectural variation is the rate of change of rivals' total output with respect to own output.

Letting $a_i$ be firm $i$'s output, $f$ the industry demand function, and $c_i$ the cost function, each firm chooses its output to maximize its profit $f(\sum_{j=1}^{n} a_j)a_i - c_i(a_i)$. Differentiating,

$$f + (1 + v_i)a_i f' - c_i' = 0, \; i = 1, \ldots, n, \tag{3.6}$$

where $v_i = \partial(\sum_{j=1, j\neq i}^{n} a_j)/\partial a_i$ is the $i^{\text{th}}$ firm's conjectural variation. If $v_i = 0$, then this is the same as condition (2.4); that is, $v_i = 0$ corresponds to the static Cournot-Nash equilibrium. In conjectural-variations models, however, other equilibria are possible. For example, if $v_i = (n-1)$ (that is each firm believes that changing its own output will induce equal changes in each of the other firms' outputs), the symmetric collusive outcome is the conjectural-variations equilibrium; if $v_i = -1$ (that is, each firm believes that changing its own output will induce an equal but opposite reaction from the other firms), the perfectly competitive outcome is the conjectural-variations equilibrium.

Although the conjectural-variations model is formally static (nowhere in the model does time enter), the model requires firms to respond to others' output changes. Thus the model is implicitly dynamic. The conjectural-variations model can be interpreted as a dynamic oligopoly model where the dynamics have been suppressed in the modelling. There are both costs and benefits to using a conjectural-variations model rather than a game-theoretic model.

The benefits come from the simplicity of the conjectural-variations model. Rather than having to formulate a fully dynamic game, it is necessary simply to solve equations of the form (3.6). The conjectural-variations model succeeds in mirroring the repeated-game model of Section 3.2 in the sense that, for example, the implicitly collusive repeated-game equilibrium can be attained as a conjectural-variations equilibrium by setting $v_i = n - 1$. It may be legitimate to use conjectural variations as a short cut if the question of how the equilibrium is maintained is not central to the modelling exercise.

The costs of using a conjectural-variations model rather than a game-theoretical model arise because of the possibility of being led into error. It is not clear what sequence of actions and reactions is hidden within the parameter $v_i$. If, realistically, reactions take time (it is not possible literally to react to a rival's simultaneous move), then that time interval might matter: what is the role here of discounting, for example? The model makes internally inconsistent assumptions: each firm makes its profit-maximizing output decision under the conjecture that there is a simple functional relationship between its rivals' outputs and the output it chooses to produce; whereas in fact the rivals are themselves solving maximization problems.

### 3.5 Stochastic games and differential games

The repeated game is the simplest kind of dynamic game. It can be generalized by adding a random element, or by having the structure of the game change over time.

Suppose that each agent's utility depends not only on all

agents' actions, but also on the outcome of some random variable. (For example, in an oligopoly game, there may be random fluctuations in demand.) Then the repeated game is converted into a *stochastic game*. An example of a stochastic game will be solved in Section 7.4.

Instead of adding a stochastic element, suppose that the structure of the game changes over time. In the repeated game, the utility obtained in any time-period depends only upon the actions taken in that period and not upon actions taken in the past; the repeated game consists simply of the repeated playing, once in each time-period, of the same static game. Suppose instead that current utility depends upon past actions. (For example, consider an oligopolistic industry extracting a natural resource. Each agent's current extraction rate may depend upon how much of the resource remains to be extracted; this in turn depends upon all agents' past extraction decisions.) In contrast to the discrete-time repeated game, games of this sort are usually modelled in continuous time (for the sake of mathematical convenience). They are called *differential games*. An example of such a game is solved in Sections 8.4 and 8.5.

The simplicity of the structure of the repeated game allows relatively complicated strategies to be considered: by conditioning their current actions on the past actions of their rivals, the agents can enforce as equilibria outcomes which all agents prefer to the static Nash equilibrium. For the sake of tractability in the more complicated stochastic games and differential games, attention is usually confined to simpler strategies, in which agents do not make their current decisions depend directly on the past actions of their rivals.

### 3.6 Cooperative solutions and noncooperative games

If a game is played repeatedly over an infinite time horizon, there can be equilibria which are Pareto optimal. In contrast to static noncooperative games, there need be no inconsistency between individual and collective rationality; pursuit of self-interest need not lead to Pareto inferior outcome.

Unfortunately, the repeated-game model's predictions are inconclusive: Pareto optimal outcomes can be equilibria, but so can the static Nash equilibrium, as well as every outcome between the static Nash point and the Pareto optima. Can the range of repeated-game equilibria be narrowed down?

It seems reasonable to suggest (although this is not supported by any rigorous argument) that, in some circumstances at least, when there are multiple equilibria, the Pareto optimal equilibria will be attained. This might be achieved through explicit negotiations taking place before the game is played; the mechanism for enforcing agreements being the retaliation possibilities inherent in the repetition of plays of the game.

Even restricting attention to Pareto optimal outcomes does not eliminate the multiplicity of equilibria: there are likely to be many Pareto optimal outcomes. Theories of cooperative games are designed to select particular Pareto optimal outcomes. This suggests a possible role for cooperative-game solutions in a noncooperative context: modelling the pregame negotiations and choosing a particular repeated-game equilibrium. If there is reason to suppose that the appropriate analysis for a particular problem involves dynamic reaction-function strategies, then cooperative solution concepts might be used to predict which Pareto optimal outcome is reached.

It was suggested in Section 2.5 that, to the extent that binding agreements cannot be made across nations, the appropriate game-theoretic model for international trade is the noncooperative game. It does not follow, however, that cooperative behavior can never arise in international relations. One law which can operate across nations is the talion: the law of retaliation in which the punishment resembles the offence committed. Thus multilateral agreements to reduce tariffs can be made to be self-enforcing by having nations retaliate to a tariff increase in violation of the agreement by increasing their own tariffs (see Section 4.6). An international cartel can discourage its members from following their short-run interests and producing more than their share of the cartel output by threatening to flood the market in retaliation (see Section 6.2). These examples suggest the relevance to international economics of the dynamic reaction-function strategies discussed in Section 3.2.

### 3.7 Bibliographic notes

On the Folk Theorem and perfect equilibria, see Aumann [7] and Rubenstein [140]. The analysis of discounted repeated games followed Friedman [44], Chapter 8; for more on the discounting case, see Abreu [1]. The concept of perfect equilibrium is due to Selten [144]. For a preliminary analysis of how agents can learn by observation about the dynamic strategies used by their rivals, see McMillan [107]. In the finite-horizon game, cooperation is an equilibrium if the definition of equilibrium is weakened to that of $\varepsilon$-equilibrium: each agent chooses a strategy which gives a level of utility within $\varepsilon$ of the largest possible utility level—see Radner [131] and Fudenberg and Levine [46]. Alternatively, in the finite-horizon game, incomplete information about the future actions of rivals can mean that an equilibrium involves cooperative behavior in some plays of the game—Kreps *et al.* [87]. Benoit and Krishna [8] proved a version of the Folk Theorem for finitely repeated games when there are multiple static equilibria.

## 4. TARIFF EQUILIBRIUM

In the 1930's, nations attempted to use their international economic policies to solve their domestic problems. These "beggar-my-neighbour" policies consisted of exchange rate depreciation, export subsidies, import tariffs, and import quotas.

Since World War II, under the General Agreement on Tariffs and Trade (GATT), nations have agreed upon several rounds of multilateral tariff reductions. It is periodically claimed, however, that the world is in danger of falling into another trade war. Consider for example the following statement by Jimmy Carter and Gerald Ford.

> Forces are at work that threaten the international trading system which has been built up since World War II and which has made possible an unparalleled expansion of trade among nations and stimulated economic growth in most parts of the globe. Governments in many parts of the world are faced with pervasive economic ills: sluggish economic growth, high rates of inflation and high levels of unemployment. In the search for solutions, they are being pressed to guarantee markets for particular interest groups by adopting protectionist measures: import restrictions, export subsidies, trade barriers in new guises. These

seductive remedies for economic ills threaten to drag the world into another period of trade war such as the one which we experienced during the Great Depression. ... To halt the drift toward economic anarchy is both a national and an international responsibility. This calls for good judgement on the part of all the countries concerned. (*New York Times*, November 21, 1982.)

What would be the effect on world trade of "economic anarchy"? Must all countries be worse off following a trade war? How can international agreements succeed in preventing trade wars? This section provides an analytic framework within which questions such as these can be posed.

### 4.1 The tariff game

A natural concept for modelling the outcome of a trade war is the noncooperative (Nash) equilibrium. Consider a game in which the players are nations and strategies are choices of tariffs. Each country seeks to maximize social welfare, which for simplicity is assumed to be a function of aggregate consumption quantities. (As Carter and Ford implied, it may be more realistic to model the government as using international trade policy to further the interests of particular groups. In effect, this would mean the less aggregated variables, such as labor income, would enter the social-welfare function which implicitly underlies the government's decision making. This issue, the political economy of tariffs, will be briefly discussed in Section 12.1.)

Consider a world with two countries, labelled 1 and 2, and two commodities, labelled 1 and 2. (The restriction to two dimensions is for simplicity of notation: the results generalize to arbitrary numbers of commodities and countries.) Suppose that each country's comparative advantage is such that, for all non-negative tariff rates, country 1 exports commodity 1 and country 2 exports commodity 2, and that whatever the changes in tariff rates, this trade pattern remains unchanged. Use subscripts to denote commodities and superscripts to denote countries. Let $p_i^j$ represent the domestic price of commodity $i$ in country $j$, and $p_i$ the world price of commodity $i$. Countries can tax international trade. Both export taxes and import taxes could be used. In the two-commodity case, a single tax suffices. Suppose therefore that each country taxes imports and not exports: country $j$ puts an *ad valorem* tariff of $t^j$ on its imports.

Country $j$ seeks to maximize the indirect utility function $u^j = V^j(p_1^j, p_2^j, m^j)$, where $m^j$ denotes disposable income. Tariff revenue is distributed in lump-sum fashion to domestic consumers. Equilibrium world prices, and therefore domestic prices, depend upon each country's choice of tariff. Thus each country's level of welfare depends upon both countries' choice of tariff.

It will be shown that, at a Nash equilibrium of this game, each country sets a tariff equal to the inverse of the elasticity of demand for its exports.

### 4.2 Nash equilibrium: diagrammatic analysis

Before solving the tariff game, it is helpful to describe diagrammatically the process of finding the Nash equilibrium.

Figure 4.1 depicts some indifference curves for country 1.

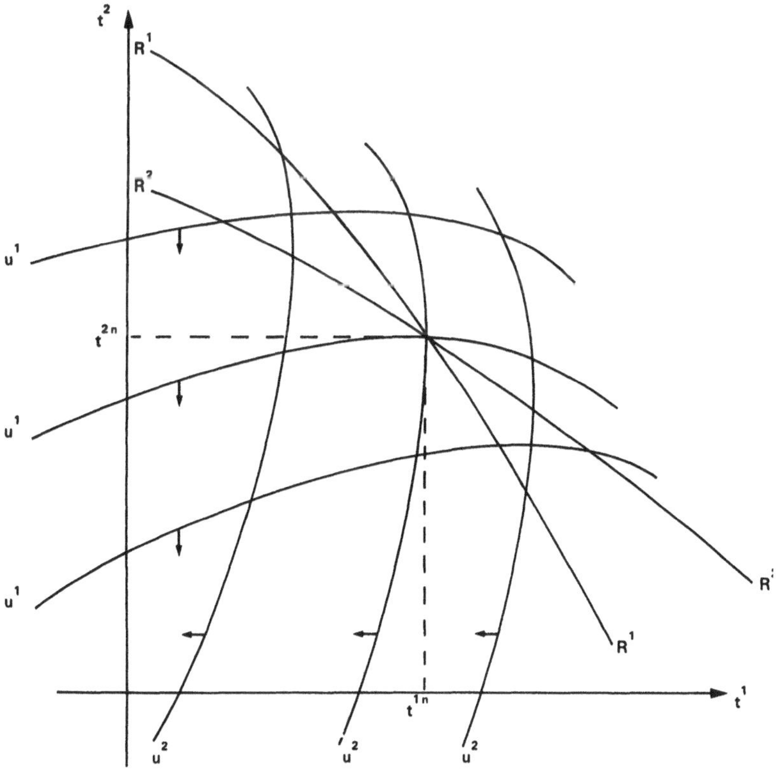


FIGURE 4.1 Tariff equilibrium.

Moving downwards across these indifference curves corresponds to increasing utility. Country 1's utility depends in part on country 2's choice of tariff; however, this is beyond the control of country 1. Given a particular choice of tariff $t^2$ by country 2, country 1 maximizes its utility by choosing a tariff $t^1$ such that the indifference curve which passes through the point $(t^1, t^2)$ is horizontal at that point. Corresponding to a different given tariff $t^2$, there is a different best reply $t^1$; the locus of country 1's best-reply tariffs is depicted as $R^1$ in Figure 4.1.

Similarly, the set of country 2's best-reply tariffs to any given tariffs of country 1, $R^2$, is constructed as the locus of points of vertical slope on country 2's indifference curves.

The Nash equilibrium embodies a mutual consistency requirement: country 1's choice of tariff maximizes its utility given the tariff of country 2, which in turn chooses its tariff to maximize its utility given country 1's tariff. The only point at which this mutual consistency is satisfied is the point of intersection in Figure 4.1 of the two best-reply curves. (This intersection need not be unique, however.) At a Nash tariff equilibrium $(t^{1N}, t^{2N})$, country 1's indifference curve has a horizontal slope and country 2's indifference curve has a vertical slope.

Notice that, in Figure 4.1, each country's Nash-equilibrium tariff is positive. Both countries are worse off at the equilibrium than they would be under free trade. Notice also why free trade is not a Nash equilibrium. Consider the free trade allocation (the origin in Figure 4.1). Is each country doing the best it can given the other country's choice of tariff? Using Figure 4.1, with country 1's tariff set at zero, country 2 can move itself to a better indifference curve by increasing its tariff above zero. Country 1 can similarly improve its welfare. It is in each country's interest unilaterally to abandon free trade; free trade is not a Nash equilibrium.

### 4.3 Nash-equilibrium tariffs

This section provides a proof of the result that the Nash equilibrium has each country setting a tariff equal to the inverse of the elasticity of demand for its exports.

The relationships among domestic and world prices are

$$\begin{aligned} p_1^1 &= p_1 \\ p_2^1 &= p_2(1+t^1) \\ p_1^2 &= p_1(1+t^2) \\ p_2^2 &= p_2 \end{aligned} \tag{4.1}$$

Let $q_i^j(p_1^j, p_2^j)$ denote the level of production of commodity $i$ in country $j$ given domestic prices in country $j$, assuming efficient production. Let $e_i^j(p_1, p_2, t^j)$ denote excess demand for commodity $i$ in country $j$, written as a function of world prices and own tariff; $e_i^j$ is therefore the quantity of commodity $i$ imported or exported by country $j$. Remembering that country 1 (2) imports good 2 (1) so that $e_i^j$, for $j \neq i$, is positive, country $j$'s tariff revenue is

$$r^j = p_i t^j e_i^j(p_1, p_2, t^j), \text{ for } i \neq j,\ i, j = 1, 2. \tag{4.2}$$

The income of consumers in country j is the sum of income from production and tariff revenue:

$$m^j = p_1^j q_1^j(p_1^j, p_2^j) + p_2^j q_2^j(p_1^j, p_2^j) + r^j. \tag{4.3}$$

Using (4.3), the indirect utility function of country $j$ can be rewritten as $u^j = H^j(p_1^j, p_2^j, r^j)$. The function $H^j$ is the indirect trade utility function, due to Woodland [164], Chapter 6. Given the standard assumptions on utility functions, $H^j$ is continuous; quasi-convex in $p_1^j, p_2^j$; homogeneous of degree zero in $p_1^j, p_2^j, r^j$; and weakly increasing in $r^j$. Moreover, using subscripts on $H$ to denote partial derivatives,

$$e_i^j(p_1, p_2, t^j) = \frac{-H_i^j}{H_3^j},\ i \neq j,\ i, j = 1, 2. \tag{4.4}$$

That is, the country's net import function is found by differentiating the indirect trade utility function $H^j$. Because $H^j$ relates utility directly to prices and tariffs, it is a convenient basis for modelling the tariff game.

General equilibrium requires that world trade is balanced; that is, each country's import demand equals the other country's export supply:

$$e_i^j = -e_i^k, j \neq k, i, j, k = 1, 2. \tag{4.5}$$

Using the relationships among world and domestic prices (4.1), the expressions for the country's income (4.2) and (4.3), and the trade-balance condition (4.5), write country 1's utility as a function of world prices, 1's tariff rate, and 2's excess-demand function:

$$u^1 = H^1(p_1, (1+t^1)p_2, -p_2 t^1 e_2^2(p_1, p_2, t^2)). \tag{4.6}$$

To find country 1's best tariff given country 2's tariff $t^2$, differentiate this with respect to $t^1$:

$$\left[p_2 + (1+t^1)\frac{\partial p_2}{\partial t^1}\right]H_2^1 - \left[\frac{\partial p_2}{\partial t^1}t^1 e_2^2 + p_2 e_2^2 + p_2 t^1 \frac{\partial e_2^2}{\partial p_2}\frac{\partial p_2}{\partial t_1}\right]H_3^1 = 0. \tag{4.7}$$

Rearrange (4.7), using (4.4) and (4.5) (using the "N" superscript to denote Nash)

$$t^{1N} = \frac{1}{\xi_2^2}. \tag{4.8}$$

Here $\xi_2^2$ is the elasticity of country 2's excess-demand function (that is, the elasticity of demand for country 1's exports):

$$\xi_i^i = \frac{e_i^i}{p_i \partial e_i^i / \partial p_i}, \; i = 1, 2. \tag{4.9}$$

Country 2's best tariff given country 1's choice of tariff is found similarly:

$$t^{2N} = \frac{1}{\xi_1^1}. \tag{4.10}$$

(Equation (4.8) defines the best-reply curve $R_1$ in Figure 4.1, and Eq. (4.10) defines $R_2$.) A Nash equilibrium in tariffs occurs when (4.8) and (4.10) are satisfied simultaneously; that is, when country 1's tariff is optimal given country 2's tariff, and simultaneously country 2's tariff is optimal given country 1's tariff.

### 4.4 Dominant equilibrium

The formulae (4.8) and (4.10) are the same as the classical optimal-tariff formula for the case of a single country without the possibility of retaliation by its trading partners (see Kemp [75], pp. 296–310). Note, however, the fundamental difference between the present problem and the standard optimal-tariff problem: in the optimal-tariff case, the elasticity $\xi_i^i$ is constant, whereas here it in general depends upon the other country's choice of tariff.

Recall from Section 2.3 that a dominant equilibrium occurs when each player's best action is independent of any other player's action. The equilibrium described by equations (4.8) and (4.10) is dominant in the special case in which each country's excess-demand elasticity $\xi_i^i$ is independent of the other country's tariff $t^j$. This is the case analyzed by Johnson [69] and Gorman [51], although they did not call it a dominant equilibrium. Johnson and Gorman showed that country $j$'s elasticity of excess demand is constant in a world of pure exchange if its utility function has the very special form

$$u^j = f^j(\alpha(e_i^j)^{\beta/\alpha} - \gamma^{\beta/\alpha}(e_j^j)^{\beta}), j \neq i, \; 1 \leqslant \beta \leqslant \alpha, \tag{4.11}$$

where $\alpha$, $\beta$ and $\gamma$ are constants.

Note another special case in which one country, though not the other, has a dominant strategy. Suppose the exports of country 1, say, make up only an insignificant fraction of country 2's consumption of good 1. Then country 1 is a perfect competitor in world trade; the demand for 1's exports is perfectly elastic. Thus country 1's best tariff rate is zero regardless of country 2's actions. This is the case analyzed in the standard optimal tariff problem.

### 4.5 Tariff reductions

Would both countries benefit if both tariffs were simultaneously reduced?

Suppose the two nations can somehow agree on a formula for mutual tariff reductions. This formula is summarized in the derivative $\partial t^1/\partial t^2$. For example, $\partial t^1/\partial t^2 = t^1/t^2$ would imply equal proportionate reductions. Differentiating expression (4.6), given

$\partial t^1/\partial t^2 > 0$,

$$\frac{\partial H^1}{\partial t^1} = \left[p_2 + (1+t^1)\frac{\partial p_2}{\partial t^1}\right]H_2^1 - \left[\frac{\partial p_2}{\partial t^1}t^1 e_2^2 + p_2 e_2^2 + p_2 t^1 \frac{\partial e_2^2}{\partial p_2}\frac{\partial p_2}{\partial t^1} + p_2 t^1 \frac{\partial e_2^2}{\partial t^2}\frac{\partial t^2}{\partial t^1}\right]H_3^1. \tag{4.12}$$

Suppose that the system is initially at a Nash equilibrium so that (4.7) holds. Then (4.12) reduces to

$$\frac{\partial H^1}{\partial t^1} = -p_2 t^1 \frac{\partial e_2^2}{\partial t^2}\frac{\partial t^2}{\partial t^1}H_3^1. \tag{4.13}$$

$H_3^1$ is the marginal utility of income, so is positive. With mutual tariff reductions, $\partial t^2/\partial t^1 > 0$. The partial derivative $\partial e_2^2/\partial t^2$ is positive if good 2 is normal. (This is because increasing country 2's tariff has two effects. It lowers the relative price of the exported good, commodity 2: this tends to increase excess demand in country 2 for good 2. Moreover, increasing the tariff tends to increase income, again increasing excess demand for good 2.) Hence $\partial H^1/\partial t^1 < 0$. Similarly, $\partial H^2/\partial t^2 < 0$. Thus any mutual tariff reductions starting from a Nash equilibrium benefit both countries.

Thus the Nash equilibrium has the property of being simultaneously individually rational and collectively irrational. Neither country can, by its own actions, make itself better off than at the Nash equilibrium. However, if some degree of cooperation can somehow be achieved, both countries can be made better off by simultaneous tariff reductions.

Although both countries gain from small mutual reductions of tariff below their Nash-equilibrium values, it does not follow that both countries would be better off if tariffs were reduced all the way to zero. (See Johnson [69] and Hamilton and Whalley [57] for some numerical examples illustrating the possibility that one country—but not both—is better off at a Nash equilibrium than under free trade. Simulations suggest that a country might gain from a tariff war if its import price elasticity is higher than the other country's.)

Consider now mutual tariff reductions from an arbitrary (not necessarily Nash) initial set of positive tariffs. Using (4.4) and (4.9), rewrite (4.12) as

$$\frac{\partial H^1}{\partial t^1} = p_2 \frac{\partial e_2^2}{\partial p_2} \frac{\partial p_2}{\partial t^1} (t^{1N} - t^1) H_3^1 - p_2 t^1 \frac{\partial e_2^2}{\partial t^2} \frac{\partial t^2}{\partial t^1} H_3^1. \quad (4.14)$$

$\partial e_2^2 / \partial p_2 < 0$ and $\partial p_2 / \partial t^1 < 0$. The second term on the right side of (4.14) is positive provided both countries simultaneously reduce their tariffs (so that $\partial t^2 / \partial t^1 > 0$). The first term on the right side is positive or negative as $t^{1N} > t^1$ or $t^{1N} < t^1$. If the current tariff $t^1$ exceeds the Nash equilibrium tariff $t^{1N}$, then country 1 must be made better off by its tariff reduction. Only if the current tariff $t^1$ is small relative to the Nash equilibrium tariff $t^{1N}$ could a country possibly be harmed by bilateral tariff reductions.

### 4.6 GATT games

It has been shown that a Nash equilibrium in tariffs has the property that, while neither country can gain by unilateral action, both countries can gain if their tariffs are reduced simultaneously. If tariff decisions were made in a cooperative-game context, any tariff changes from which both countries gained could be achieved. There is, however, no international legal system to enforce agreements between countries. This is a game without the possibility of binding agreements: it is a noncooperative game.

Since the Second World War, multilateral agreements under the aegis of the General Agreement on Trade and Tariffs have significantly reduced tariffs. Hamilton and Whalley [57] computed Nash equilibria in a general equilibrium model. Estimated tariff rates were in excess of fifty percent for the United States and the European Economic Community. This is much higher than current tariff rates; it is, however, the order of magnitude of tariff rates that ruled after the tariff wars of the 1930's. This suggests that, since the Second World War, GATT has succeeded in reducing tariffs far below their Nash equilibrium levels.

Given that the game is noncooperative, how can GATT succeed in generating multilateral tariff reductions? The answer is provided by Dam [31], p. 79:

> The essence of the GATT system lies not in the abstract legal relationships created by a tariff concession but rather in the enforcement mechanism. Under Article XXIII the principal sanction for an increase of a duty in violation of a binding is the suspension by interested contracting parties of concessions made to the offending contracting party. There is no punitive sanction for nonperformance of the promise implicit in a tariff concession. The consequence of nonperformance is thus merely the reestablishment, at the option of an interested party and subject to the approval of the contracting parties, of the preexisting situation (although the retaliatory suspension may be on items not originally negotiated with the offending contracting party).

(In the quaint jargon of GATT, a tariff "concession" is a tariff reduction, and a "binding" is a commitment not to raise a tariff.) For example, in early 1984, the European Economic Community retaliated to the United States' imposition of tariffs and quotas on imports of European specialty steels by imposing quotas on a long list of American exports including plastics and steels.

The game modelled above is static: only current consumption matters. An alternative way of modelling the tariff-setting game is in dynamic terms, as a repeated game. One of the equilibria of a repeated game is simply the repeated playing of the static-game equilibrium. There are, however, other equilibria, which are Pareto superior. As was shown in Section 3.2, provided neither country discounts future consumption too heavily, a situation which both countries prefer to the static equilibrium can be supported as a dynamic equilibrium by each country following a strategy of punishing other countries' tariff increases by increasing its own tariffs.

According to the quotation from Dam, GATT enforces agreements by having countries respond to violations by increasing their own tariffs. This suggests a repeated-game interpretation of GATT's role. There are many repeated-game equilibria. GATT's function can be thought of as initiating the playing by each nation of some particular dynamic strategy which results in an equilibrium Pareto-superior to the static equilibrium. Once a Pareto-superior equilibrium is attained, it is by definition self-enforcing. In addition to selecting which Pareto-superior equilibrium is to be achieved (recall from Section 3.2 that there are many such potential equilibria) and arranging that each nation plays the dynamic strategy consistent with that equilibrium, GATT's role is to ensure that retaliation to violations of agreements is kept within prescribed bounds so that it does not drive the system permanently back to the static equilibrium (compare with Dam [31], pp. 80–81).

### 4.7 Bibliographic notes

Early discussions of the question of tariff retaliation were provided by Kaldor [71], Scitovsky [142], and Robinson [136]. Johnson [69] solved the tariff-retaliation problem in some special cases. Gorman [51] analyzed what was described in Section 4.4 as the dominant-equilibrium case. Panchamukhi [126] pointed out the game-theoretic nature of the problem. Kuga [90] and Otani [123, 124] gave proofs of the existence of a general equilibrium with strategic tariff-setting. Variations on the theory of tariff retaliation were provided by Horwell [64], who considered specific rather than *ad valorem* tariffs; Rodriguez [139] and Tower [156], who analyzed quota retaliation (and showed that there is an important difference from the tariff case: at a quota equilibrium, trade is eliminated); Weymark [162], who modelled each country as using tariff revenue, rather than tariff rate, as it decision variable; Riezman [134], who modelled the tariff game as a prisoners' dilemma; and Thursby and Jensen [155], who presented a conjectural-variations model. Hamilton and Whalley [57] provided numerical estimates of Nash-equilibrium tariff rates.

Mayer [104] discussed negotiated tariff reductions starting from the Nash equilibrium. Markusen [102] considered the distributional effects of tariff reductions from arbitrary initial tariff rates. Allen [3] used the Shapley value to model GATT negotiations. GATT as a legal entity was discussed by Dam [31].

The related problem of trade warfare—in which one country uses tariffs, embargoes, etc., to reduce the welfare of another country—was analyzed by Wan [160]. (Parts of this are summarized in Kemp [73], Chapter 15.) On the role of game theory in analyzing international conflicts, see Harsanyi [60].

## 5. TARIFFS AND FOREIGN COMPETITION

What is the appropriate government action if firms, either foreign or domestic, have the power to affect the price in the domestic market? It is sometimes claimed that the most effective way to restrain a domestic monopoly is to allow foreign firms to compete with it. A clear statement of this opinion was given by Friedman and Friedman [45], p. 76:

> The extent of competition at home is closely related to international trade arrangements. A public outcry against 'trusts' and 'monopolies' in the late nineteenth century led to the establishment of the Interstate Commerce Commission and the adoption of the Sherman Anti-Trust Law, later supplemented by many other legislative actions to promote competition. These measures have had very mixed effects. They have contributed in some ways to increased competition, but in others they have had perverse effects.
>
> But no such measure, even if it lived up to every expectation of its sponsors, could do as much to assure effective competition as the elimination of all barriers to international trade. The existence of only three major automobile producers in the United States— and one of those on the verge of bankruptcy—does raise a threat of monopoly pricing. But let the automobile producers of the world compete with General Motors, Ford, and Chrysler for the custom of the American buyer, and the specter of monopoly pricing disappears.

Arguments of this kind have influenced public policy towards big business. In antitrust cases, United States courts have increasingly upheld, in defence of firms accused of having too large a share of domestic production, arguments about the extent of competition from foreign firms. The Canadian Combines Investigation Act has a provision for selectively reducing tariffs to correct a monopolistic distortion.

Friedman and Friedman pointed out one way in which lowering tariffs affects a country's welfare: the increased competition lowers domestic prices and therefore makes consumers better off. If, however, the international market is oligopolistic, then there are two additional consequences of lowering tariffs, which tend in contrast to lower the country's welfare. First, the market power of the foreign producers means that the importing country pays a price greater than marginal cost. By imposing a tariff, the country can extract some of the monopoly surplus. Second, the protected domestic monopoly earns profits. Some of these profits go to overseas firms after the lowering of tariffs, tending to lower the country's welfare. This section develops a model which weighs these three consequences of changing tariff levels when producers have oligopoly power in world trade.

### 5.1 Foreign and domestic competition

Consider a simple model in which a commodity is both produced within the country by a single domestic firm and imported from a single foreign firm. Assume that, perhaps because of prohibitive foreign tariffs, this country's market is isolated from the rest of the world, so that the price within the country is determined by domestic demand. What is the country's best tariff policy in response to the market power of both the foreign producer and the domestic producer?

Consider a game with three agents: the two firms plus the government. Suppose the government plays as the Stackelberg leader, and the two firms play as Stackelberg followers. The leader commits himself to an action before the followers. Knowing all the parameters of the game, the leader can predict the responses of the followers; this knowledge is used in deciding his own action. To solve for the Stackelberg equilibrium, the first step is to find the firms' output responses to any choice of tariff by the government.

Denote by $x$ the quantity demanded within the country, and by $p(x)$ the country's inverse demand curve: $p'(x) < 0$. Denote by $x^d$ and $x^f$ the sales within this country of the domestic firm and the foreign firm, respectively: $x^d + x^f = x$. Suppose that each firm produces under constant average cost, denoted by $c^d$ and $c^f$. (Note that if average cost varied with output, the decision on how much to sell in any one country would depend on how much is being sold elsewhere; with constant average costs, these decisions across different countries can be made separately.) If a tariff of $t$ is levied on imports, the foreign firm's profit is

$$\pi^f = x^f[p(x) - t] - c^f x^f \tag{5.1}$$

and the domestic firm's profit is

$$\pi^d = x^d p(x) - c^d x^d. \tag{5.2}$$

With the tariff rate given, the Cournot-Nash equilibrium of the game between the two firms alone is found (assuming an interior solution) by solving simultaneously:

$$p'(x)x^f + p(x) - t - c^f = 0 \tag{5.3}$$

$$p'(x)x^d + p(x) - c^d = 0. \tag{5.4}$$

Consider now the effect on the firms' equilibrium of a change in the tariff rate. Differentiating (5.3) and (5.4) totally with respect to $t$ yields (using $x_t^d$ to denote $dx^d/dt$, etc.)

$$x_t^d = \frac{-p''x^d - p'}{p'(p''x + 3p')} \tag{5.5}$$

$$x_t^f = \frac{p''x^d + 2p'}{p'(p''x + 3p')}. \tag{5.6}$$

In order to find the effects of the tariff on domestic and foreign production, that is, to attach signs to $x_t^d$ and $x_t^f$, assume that

$$x^d p'' + p' \leqslant 0 \tag{5.7}$$

and

$$x^f p'' + p' \leqslant 0. \tag{5.8}$$

Condition (5.7) says that the domestic firm's marginal revenue decreases as the foreign firm's output increases; similarly (5.8) says that the foreign firm's marginal revenue is decreasing in the domestic firm's output. The rationale for postulating (5.7) and (5.8) is that Novshek [118] showed that, along with some innocuous technical conditions, these are sufficient to guarantee existence of a Nash equilibrium for this problem. Condition (5.7) ensures that the numerator of the right side of (5.5) is nonnegative, and, given $p' < 0$, the numerator of the right side of (5.6) is nonpositive. Adding (5.7) and (5.8), and using $p' < 0$, shows that the denominators of the right sides of (5.5) and (5.6) are nonnegative. Hence $x_t^f \leqslant 0$ and $x_t^d \geqslant 0$: increasing the tariff reduces the output of the foreign firm and increases the output of the domestic firm. Which effect is bigger? Adding together (5.5) and (5.6) gives

$$x_t = \frac{1}{p''x + 3p'} < 0. \tag{5.9}$$

Thus increasing the tariff has the net effect of lowering total domestic consumption; the decrease in foreign sales outweighs the increase in domestic sales.

The behavior of the Stackelberg followers has been characterized. Consider now the Government, behaving as Stackelberg leader, choosing the tariff while taking into account the actions of the follower-firms as summarized by equations (5.3), (5.4), (5.5) and (5.6). Assuming that this commodity represents a small fraction of the country's total consumption, the gross benefit to domestic consumers is the area under the demand curve:

$$b(x) = \int_0^x p(z)\, dz. \tag{5.10}$$

The consumer surplus is therefore $b(x) - xp(x)$. The country's total welfare is the sum of consumer surplus, tariff revenue, and domestic profits:

$$u = b(x) - xp(x) + tx^f + x^d p(x) - c^d x^d. \tag{5.11}$$

This expression exhibits the three welfare consequences of changing tariffs already mentioned. Since lowering the tariff level increases total domestic consumption, consumers are made better off; however, when the tariff is lowered, tariff proceeds decrease and domestic profits decrease. Thus whether lowering the tariff improves the country's welfare depends on the parameters and on the initial level of the tariff. (This is a second-best problem: the oligopoly power of the domestic and foreign firms is taken as given.)

The government's Stackelberg-leader action is found by differentiating the welfare function (5.11) with respect to $t$, using (5.10), and solving for $t$:

$$t^s = \frac{(p'x_t - 1)x^f - (p - c^d)x_t^d}{x_t}. \tag{5.12}$$

Adding together (5.7) and (5.8) gives $p''x + 2p' \leqslant 0$. Using this, (5.9) reduces to $p'x_t \leqslant 1$. From this, plus $x_t^d > 0$, $x_t < 0$, and $p \geqslant c^d$, the Stackelberg-equilibrium tariff is positive. If the tariff is initially too low, then an increase in the tariff will cause an increase in rents accruing to the country which is larger than the decrease in consumer surplus.

Thus, although the tariff causes a loss of consumer surplus, the oligopolistic nature of the market means that the tariff should not be eliminated.

### 5.2 Industrial organization and trade

This model could be extended in a number of ways to examine questions on the borderline between industrial organization and international economics. One straightforward extension is to allow any finite number of firms at home and abroad; this would not change the results in any essential way. This raises the further question of entry of new firms into the industry. Attention need not be confined to tariffs; the efficacy of other government policies such as imposing quotas, subsidizing domestic producers, or encouraging or discouraging mergers of domestic firms could be considered. By adding a foreign country to the model, the spillover effects of one country's policies on the equilibrium in the other country could be investigated. By changing the cost functions to allow economies of scale, the model could be used to consider the effect of the size of the domestic market on a country's international competitiveness.

Note that the result that the tariff should be positive depends upon the firms' ability to affect the domestic price. If the world market were perfectly competitive, so that the country faced an infinitely elastic foreign supply curve, the country should eliminate its tariff. Thus the industrial-organization effects of tariffs described above will not occur in the small, open economy of classical trade theory.

### 5.3 Monopoly in an exhaustible resource

The idea that a tariff can be used to extract surplus from a foreign monopolist is most clearly illustrated by the case of trade in an exhaustible resource such as oil.

Models of exhaustible resources typically assume that the resource is eventually exhausted. While this seems a plausible assumption, it unduly restricts the scope of actions available to a supplier of the resource: it may be in a supplier's interest to hold some of the resource off the market. To see this, modify the model of Section 5.1 so that the foreign firm has a fixed total stock $X$ of the exhaustible resource, which can be extracted at constant marginal cost $c$; for simplicity assume there are nondomestic reserves and so no domestic production. Retain the static model, in effect assuming that the world lasts only one period. (Using a static rather than a dynamic model assumes away many of the important questions of resource economics; however, the one-period model is rich enough to address the question of the distribution of monopoly rents. An example of a dynamic game of resource exploitation will be analyzed in Section 8.) The model of Section 5.1 was consistent with the interpretation that there were several importing countries, but because of the absence of arbitrage and the constant-returns-to-scale production, decisions about sales in each country could be taken separately. For the present analysis, however, suppose there is only one importing country.

Given the fixed total stock of the resource, the firm's profit-maximum condition (5.3) must be rewritten as, for $x \leqslant X$,

$$p'(x)x + p(x) - t - c \geqslant 0, \text{ with} = \text{only if } x < X. \tag{5.13}$$

(Since there is only one supplier now, the $f$ and $d$ superscripts can be dropped.)

The model is one of bilateral monopoly: the single consuming country buys from the single supplier. It is sometimes claimed that the fact that the stock of the resource is fixed means that the importing country can set a tariff high enough to extract almost all the monopoly rents of the supplier. This ignores, however, the monopolist supplier's ability to leave some of the resource unextracted. If the resources were inelastically supplied, a tariff could be used to extract all of the rents; however, the total stock of the resource being fixed does not imply the resource is supplied inelastically.

Solving (5.13) as an equation, the highest tariff under which the supplier will extract all of the resource is

$$t^* = p(X) + p'(X)X - c \tag{5.14}$$

From (5.1), the supplier's profit is then

$$\pi = -X^2 p'(X) > 0. \tag{5.15}$$

Because the demand curve is downward sloping, the importing country cannot extract all of the monopoly rent. If the tariff were set high enough to absorb all the rent in the event of the resource being completely exhausted, then the supplier could earn positive profit by supplying less than the full amount of the resource. In other words, for a tariff level higher than that given by (5.14), the supplier's marginal revenue is less than marginal cost at the point of complete extraction.

The Stackelberg equilibrium of this game either has the leader-government setting its tariff according to (5.14) and the follower-firm extracting the entire stock of the resource (a corner solution), or the government setting its tariff as in Section 5.1 and the supplier extracting only a part of the stock (an interior solution). In either case, the supplier earns positive profits despite the fact that the size of the stock is fixed. The importing country's tariff extracts some, but not all, of the monopoly rents.

### 5.4 General equilibrium

Traditionally, the theory of international trade is set in a general-equilibrium framework. The model developed above is partial equilibrium; it ignores income effects, substitution across commodities, and changes in factor prices. The use of consumer surplus in the country's welfare function is approximately valid if expenditure on the commodity is a small fraction of gross domestic expenditure. To consider income and cross-substitution effects and changes in factor prices, and to allow the possibility that expenditure on the commodity might be relatively large, the analysis would have to be recast in general-equilibrium terms. This is, however, not easy to do, and should not be done simply by reinterpreting the variables. When interactions across markets are fully recognized, equilibrium may not exist. The question of existence of general equilibrium is no mere technicality: it is a question about the internal consistency of the model. Strategic behavior in a multi-market context may preclude existence of equilibrium: Roberts and Sonnenschein [135] gave some apparently reasonable examples of oligopoly models with nonexistent general equilibrium.

### 5.5 Bibliographic notes

The analysis of Section 5.1 was based in part on the models of Brander and Spencer [19] and Dixit [34]. For more on the welfare effects of tariffs and subsidies when international markets are oligopolistic, see Dixit and Grossman [35] and Eaton and Grossman [40].

Policies toward mergers when firms face foreign competition were analyzed by Ordover and Willig [122] and Dixit [34]. Brander [16], Hwang [65], Krugman [89], and Shaked and Sutton [145] investigated the gains from international trade when oligopolistic firms produce under economics of sale. Brander and Spencer [18] and Dixit and Kyle [36] modelled the interactions between tariff policy and entry of firms into an industry. Horstmann and Markusen [63] modelled the interactions between multinational and domestic firms and discussed the appropriate host-country commercial policies. Porter [128] investigated the consequences of tariffs in a spatial economy. Krishna [88] argued that import quotas could help firms to maintain a collusive equilibrium. Spencer and Brander [149] modelled international competition in research and development. Harris and Cox [59] estimated the gains to Canada from the increased competition consequent on tariff reductions.

That an importing country can choose its tariff so as to extract monopoly rent from an exporter of an exhaustible resource was shown by Kemp and Long [78] and Bergstrom [9], see also Brander and Djajic [17] and Lawrence and Levy [91]. Maskin and Newbery [99] showed that the importing country's inability to commit itself to future tariffs might make it worse off than in the absence of tariffs. Other strategic variables available to a country buying an exhaustible resource include domestic production (Lee [92], Robson [138]), stockpiling (Nicholls and Zeckhauser [117]), capital accumulation (Chiarella [23], Ulph and Ulph [157]), and the development of a substitute (Dasgupta, Gilbert and Stiglitz [32], Gallini, Lewis and Ware [47]). A cooperative-game model of the defensive reactions of a nation buying a exhaustible resource to an embargo threat was developed by Crawford, Sobel, and Takahashi [30].

General-equilibrium models of international oligopoly (assuming constant-elasticity-of-substitution utility functions) were developed by Melvin and Warne [110] and Markusen [103]. Proofs of existence of general equilibrium in the presence of Cournot-Nash oligopoly were provided by Negishi [114] and Arrow and Hahn [6], Chapter 6; however, as Roberts and Sonnenschein [135] pointed out, these proofs rely on *ad hoc* assumptions about the nature of demand.

## 6. INTERNATIONAL CARTELS

The earliest and most frequently studied game in economics is the oligopoly game. In this game is displayed the archetypical dilemma which arises in noncooperative games: there are outcomes which the participants unanimously prefer to the static equilibrium, but problems of coordination must somehow be resolved before such preferred outcomes can be attained. Collusion, if successful, maximizes the profit of each producer in an oligopolistic industry; but each has an immediate incentive to produce more than its share of the collusive output level.

The history of international cartels is an interesting source of case studies of the playing of oligopoly games in practice, illustrating the tension between collusion and competition. Unlike domestic cartels, information on the behavior of international cartels is relatively freely available. The absence of international antitrust laws means that international cartels need not operate covertly. Because cartels within particular countries are often illegal, their operations are conducted in secrecy; knowledge about domestic cartels is biased because in most cases it is only the unsuccessful ones which are observed.

In marked contrast to the trade theorist's assumption of universal perfect competition, cartels are common in international trade. In addition to the conspicuous example of the Organization of Petroleum Exporting Countries, cartels have at some time been formed by producers of the following commodities: rubber, mercury, tin, sugar, steel, tea, copper, tropical timber, sulphur, potash, phosphate rock, magnesium, iodine, diamonds, wheat, platinum, coffee, bananas, sodium sulphate, sodium nitrate, sisal, cocoa, bauxite, iron ore, tantalum, nickel, tungsten, cobalt, zinc, lead, cement, and uranium. Nor has collusion been confined to primary commodities: there have been international cartels in air transport, shipping, chemical products, and electrical goods.

This section summarizes some facts from the history of international cartels, and discusses them in the light of the theory of repeated games.

### 6.1 Repeated games and cartels

Interpret the repeated game of Section 3.2 as a model of oligopoly, the $a_i$'s representing outputs and the $u^i$'s profit functions. Provided the producers do not discount future profits too heavily, outcomes on the profit-possibility frontier can be supported as dynamic equilibria if the individual producers discourage fellow cartel members from producing too much by threatening retaliatory large outputs. Hence a cartel can be self-enforcing in a dynamic model.

The larger is the right-hand side of inequality (3.5), the more likely it is that firm $i$ will produce its cartel level of output $a_i^*$. There are two reasons for this. First, the larger it is, the more likely it is that, for a given discount rate $\alpha_i$, inequality (3.5) holds. Second, even if the discount rate is close enough to one that (3.5) holds, the larger is the right-hand side, the more likely it is that the costs of negotiating an agreement and the costs of observing fellow cartel-members' outputs are outweighed by the benefits from the cartel's existence. (Neither of these costs is explicitly modelled.) The right-hand side of (3.5) is larger the more producer $i$'s share of the cartel's profit exceeds its static (Cournot) equilibrium profit; and it is larger the smaller is the increase in the producer's profit if it deviates from its cartel output.

While the analysis of Section 3.2 shows that it is possible for a cartel to be self-enforcing, it also points out a number of difficulties which a successful cartel must overcome. First, the repeated game has many equilibria other than cartel outcomes. The industry must somehow find its way to an outcome on the profit-possibility frontier. Second, there are many outcomes on the profit-possibility frontier, each with a different distributions of profits among the cartel members. The successful cartel must somehow resolve conflicts among its members over market shares.

### 6.2 International cartels in practise

International cartels are usually initiated by explicit negotiations among the producers. The cartel arrangements range from loosely defined agreements to restrict output to formal arrangements providing centralized machinery for allocating output quotas and fixing prices, imposing fines on members who deviate from their quotas. Some have only a slight or short-lived effect on market prices. Others succeed in eliminating all competition among the members.

A common way in which a cartel controls sales is by a geographical division of the world market. Each producer is free to set its own price and output within its own geographical area, but may not, according to the terms of the cartel agreement, sell in other members' areas. In other cartels, members sell worldwide but sell no more than a preassigned amount at a preassigned price. More rarely, a cartel may have a central sales organization which pools orders and allocates them across members.

The following data are drawn from a survey of the econometric and anecdotal literature on international cartels by Eckbo [41], Chapter 3. This evidence covers 51 cartel agreements in 18 industries, the earliest agreement being in 1819 and the latest in 1964.

A test of the success of the cartel is whether it achieved a price close to the monopoly price. In the absence of detailed data on cost and demand functions, Eckbo defined a cartel as being successful if it raised the price to at least three times the marginal production cost of the highest-cost cartel member. On this criterion, only 19 of the sample of 51 cartels (or 37 percent) were successful.

The cartels did not seem to last for a long period of time. One outlier, the iodine cartel established in 1878, lasted for 61 years. Of the remaining cartels which were successful in raising price to three times marginal cost, the length of life ranged from 2 to 18 years, with a median of 5 years and a mean of 6.6 years.

Since success was not long-lived, the question of why the cartels broke down is important. Of the cartels which successfully raised the price, 3 (out of 9 for which such information was available) broke down for nonmarket reasons, such as government intervention or war. Of those successful cartels which broke down for market-related reasons, 7 out of 16 (44 percent) broke down because of internal conflict among the cartel members; the remaining 9 out of 16 (56 percent) broke down because of external forces (usually competition from nonmember suppliers, but occasionally retaliation by buyers).

Those industries in which a successful cartel was established but later broke down were often able to re-establish the cartel a few years after the breakdown.

High market shares facilitate collusion. Defining an industry to be concentrated if the 4 largest producers produce more than 50 percent of the total output of the industry, 15 out of 19 (79 percent) of the successful cartels were concentrated. In 14 out of the 19 cases (74 percent), the cartel members' share of total production in the industry was above 75 percent.

Characteristics of the demand function affect cartelization. Inelastic demand obviously facilitates a price rise from an initial noncartel outcome; 7 out of the 9 successful cartels for which demand data were available faced a price elasticity of demand of less than one. One of the determinants of demand elasticity is the availability of substitutes: in 8 of the 9 cases, no short-term substitutes for the commodity were available outside the cartel. However, in 7 of the 9 cases there were long-term substitutes, suggesting an explanation for the observation that a common reason for the eventual breakdown of successful cartels was external competition. In all 9 cases the income elasticity of demand exceeded one.

Small cost differences among the producers make it more likely that agreement will be reached over the appropriate price for the cartel to set. In 8 of 9 cases of successful cartels, the cost differences among cartel members were less than 50 percent.

Similar conclusions can be drawn from a study of cartels of the inter-war period by Lovasky [95]. There is considerable overlap between the Eckbo and Lovasy samples.) Of 22 raw-material cartels, 9 were judged by Lovasy to have successfully raised prices, 8 were unsuccessful, and no price information was given for the other 5. Of the 9 successful cartels, the shortest-lived lasted 4 years, and the longest-lived was still in existence after 24 years. Growing competition from outsiders and internal conflict over market shares appeared to be equally common sources of cartel breakdown. Many of the industries which attempted cartelization had excess capacity. The success of a cartel in raising prices often stimulated new producers to enter the industry, destroying the cartel.

### 6.3 Cartel history and repeated games

How do these observations on the determinants of cartel success relate to the cartel theory sketched in Section 6.1?

The success of a cartel depends first on its ability to be self-enforcing; second on agreement being reached on the division of the market among members; and third on the extent of external competition.

The cartel can be self-enforcing if the inequality (3.5) holds. This is more likely if demand is inelastic; then industry profit at the cartel point is much higher that at the static Cournot-Nash equilibrium, so that the numerator of (3.5) (each producer's opportunity cost of deviating from its cartel output) is large. The larger is a producer's market share, the more will increasing its own output lower the market price; that is, the proportionately smaller is the denominator of (3.5) (the producer's short-run gain from deviating from its cartel output). Inequality (3.5) presumes that cartel members are able to increase output in retaliation to a deviation: thus a successful cartel should have excess capacity. These three features of the theory are borne out by the observations reported in Section 6.2. There is a fourth testable prediction of the theory, on which observations do not seem to be available. Cartel members weigh the long-run costs of deviating against the immediate gains. This intertemporal comparison is affected by changes in the interest rate: the model predicts cartel breakdowns to be more frequent when interest rates are high.

For deterrent strategies to be workable, it is necessary for members to be able to observe other members' output levels. This is less costly the fewer the members of the cartel: hence this provides a second reason why concentration aids the cartel's success.

That cartel arrangements were frequently re-established some time after a breakdown is consistent with the cartel members retaliating to other members' deviations by increasing their own outputs, but not persisting indefinitely with this punishment.

The second problem for a cartel is to reach agreement on the division of the market. As noted in Section 6.1 there are many profit-possibility frontier outcomes. That cartels were observed frequently to break down because of internal conflicts indicates the force of conflicts over the distribution of profits among cartel members. Evidently this issue is addressed during the pre-cartel negotiations among the producers. This suggests a third reason why industry concentration aids cartelization: the fewer the participants in the negotiations over the division of the market, the more likely an agreement will be reached. The smaller the differences in production costs, the less diverse are the interests of the producers with respect to how high the cartel price should be and therefore the more likely the negotiations are to be concluded successfully. Recurrent disagreements over market shares are presumably less severe if total demand is increasing. Hence the observed association between income-elastic demand and successful cartels: if demand is income-elastic, demand increases faster than world economic growth.

The third problem for cartels is outside competition. The significance of this problem is indicated by the prevalence of cartel breakdowns because of external forces. The smaller is a potential cartel's market share, the less the cartel can affect the price and the smaller are the potential gains from cartelization. If new producers enter the industry, the members of an established cartel must negotiate a new equilibrium point, with lower profits for the members. Thus the possibility of impasse in bargaining results in the possibility that any new entrant will destroy the collusive arrangement. The more successful a cartel is, the more susceptible it is to external competition, because the high cartel prices may make it profitable for alternative producers to incur the fixed costs of exploration or research and development which precede entry into the industry; hence the observed longrun availability of substitutes and the observation that cartels usually survive for only a few years.

### 6.4 Bibliographic notes

The repeated-game oligopoly theory of Section 6.1 was based on Friedman [44], Chapters 2, 8. On the role of observability of firms' actions in repeated-game equilibria, see Green [53], Porter [128], and Robson and McMillan [138]. The pricing and output-share decisions of OPEC were modelled as a cooperative game by Hnyilicza and Pindyck [62] and Gately [49].

The data on international cartels were drawn from Eckbo [41], Chapter 3. For case studies of international cartels, see Lovasy [95], Stocking and Watkins [153], and Pindyck [127].

## 7. PRICE STABILIZATION

The Report of the Independent Commission on International Development Issues (the Brandt Commission) argued strongly for a redistribution of world welfare toward the less-developed countries. Stabilizing world prices of primary commodities is one of the ways the Commission suggested that welfare in less-developed countries could be improved. Most of the Third World's exports are primary commodities; moreover, many countries are dependent for export earnings on one or two commodities. Commodity prices tend to be volatile. There are fluctuations in demand because of business cycles in the developed countries and because of changes in buyers' stockpiling policies. There are fluctuations in supply because of the variability of harvests. The Brandt Commission argued that: "Action for the stabilization of commodity prices at renumerative levels should be undertaken as a matter of urgency" (Brandt, *et al.* [20], p. 158). Historically however, attempts (some sponsored by the United Nations) to secure international commodity agreements to stabilize prices by means of internationally managed buffer stocks have usually not been successful. The Brandt Commission therefore suggested that:

> Where international agreements have not yet been established or where they have proved difficult to establish, the availability of finance for national stocking could be beneficial to producing countries in discouraging selling at inappropriate times. Some countries have already undertaken such stocking on their own but poorer countries are unable to afford it. The provision of such finance in particular circumstances, especially to poorer countries, would give further encouragement to national stocking whether through internal or external financing, since greater benefits would accrue to all concerned from the greater price support that would result [20, pp. 152–153].

Many world primary-commodity markets are imperfectly competitive: individual exporting countries often have some market power. For example, Brazil has a 42 percent share, and Tanzania a 28 percent share, of world trade in sisal; although market power does not depend only on market share, this would appear to be a classic oligopolistic market. Malaysia's exports of palm oil are 69 percent, of copra are 55 percent, of rubber are 51 percent, and of tin are 36 percent of the amounts traded internationally (Newbery [115]). Thus models of price stabilization which are not game theoretic and which therefore ignore oligopolistic interactions fail to capture an important feature of international commodity markets.

Under what circumstances is price stabilization in the interests of the producing countries? Could it be achieved, as the Brandt Commission suggested, by individual countries acting independently? This section addresses these questions via a stochastic-game model of the use of stockpiling by exporting countries to stabilize world prices.

### 7.1 Fluctuations in world demand

The model to be developed is a partial-equilibrium model of storage and pricing of a particular commodity.

Suppose there are several countries producing this commodity, and that within each country there is centralized control over exports, perhaps through a national marketing board.

Assume that each country's product is somewhat differentiated from any other country's product. (This is for a technical reason: it rules out the discontinuities which would arise if the products were homogeneous; these discontinuities being caused by the sudden switching of buyers from one producer to another if the latter's price drops slightly below the former's.) Because the products are differentiated, the demand function facing each country is negatively sloped. The demand for each country's product depends on all countries' prices.

There is evidence that demand fluctuations are more important than supply fluctuations in generating the volatility of less-developed countries' export markets (Nurkse [119]). (Short-run fluctuations in supply are important for some crops, but presumably are less important for minerals.) Assume therefore that demand fluctuates randomly and that cost functions are nonstochastic.

Because demand functions are not infinitely elastic, each country is able to choose a price to charge. Because cost functions are nonstochastic, each country can also choose its output level. Assume that, before it observes the current level of demand, each country must commit itself to both a price and an output level, and that this price-output pair remains in force for a specified period of time ("supply may not be able to adjust readily; for example, crops may already have been planted, and changing mineral production capacity takes time"—Brandt, *et al.* [20], p. 146).

Because both price and output are decided in advance of each period's demand, supply will equal demand only by chance. Each country meets discrepancies between supply and demand by building up or drawing down its stockpile of the commodity. The holding of stocks incurs costs (interest costs plus the physical costs of storage).

Decisions about price and production are made repeatedly, at the start of each time-period in a dynamic discrete-time model. Each country seeks to maximize the expected discounted value of the stream of rents generated by producing and selling the commodity.

When is it in each country's interest, taking into account the actions of the other countries selling this commodity, to maintain a price constant over time by adding to its stockpile of the commodity when world demand is low and drawing down its stockpile when world demand is high?

### 7.2 Stochastic games

The game sketched in Section 7.1 is an example of a stochastic game. A stochastic game is a repeated game with a random structure: the game which is actually played in any particular time period is drawn according to a known probability distribution from some set of constituent games. (The repeated game of Section 3.1 is a degenerate stochastic game in which the set of constituent games has only one element.)

The dynamics of the game are determined by the agents' decision rules and by the probability distribution function which determines the selection of each period's game.

Each agent attempts to maximize the expected present value of utility. At Nash equilibrium, each agent's expected discounted utility is at a maximum given the decision rules of all the other agents.

Solving such games is in general difficult. In some circumstances, however, a stochastic game can be shown to be equivalent to a nonstochastic repeated game of that sort examined in Section 3.1: one of the equilibria then is simply the repeated playing of the Nash equilibrium of the component static game. Such an equilibrium was called a *myoptic equilibrium* by Sobel [148].

### 7.3 Prices, outputs, and storage

Denote the number of countries by $n$; $n$ may be, but not need be, a large number. The countries need not be of equal size: some may have a large share, and others an insignificant share, of world trade. Because of the assumption of product differentiation, each country, no matter how small its share of world trade, faces a negatively sloped demand curve.

Each country's sales depend not only on the price it decides to charge, but also on the prices charged by all the other countries. In addition, each country's demand curve fluctuates randomly. Let $x_j^t$ denote the demand facing country $j$ in period $t$; $x_j^t$ is a random variable whose distribution depends on the prices charged by all countries in period $t$ but not on the value of any variable in the past.

At the beginning of each period, and before the current

realization of demand can be observed, each country decides its price and output for the duration of that period. Denote by $p_j^t$ and $q_j^t$ the price and output of country $j$ in period $t$.

Let $z_j^t$ denote the amount of the commodity country $j$ has in storage at the start of period $t$ (carried over from the previous period). Denote the costs of storage by $I_j(z_j^t)$; storage costs can be thought of as being incurred at the end of a period.

Denote by $y_j^t$ the total amount of the commodity available to satisfy demand for country $j$'s output in period $t$; $y_j^t$ is the sum of the amount produced this period and the amount carried over in storage from the previous period:

$$y_j^t = z_j^t + q_j^t. \tag{7.1}$$

$y_j^t$ is country $j$'s supply in period $t$.

Stocks carried over to period $t+1$ are the difference between supply and demand in period $t$:

$$z_j^{t+1} = y_j^t - x_j^t. \tag{7.2}$$

Notice a potential difficulty here: there is nothing in the model to prevent the stock $z_j^{t+1}$ from being negative. This difficulty can be resolved in one of two ways. If it is possible for a country to meet excess demand by promising future delivery, then there is nothing inconsistent in $z_j^t$ being negative: a negative value of $z_j^t$ represents unfilled orders. (One of the costs of holding stocks included in $I_j(z_j^t)$ for negative $z_j^t$ might then reflect the risk that some customers will buy elsewhere rather than wait.) Alternatively, if the country loses sales whenever it cannot satisfy demand, so that $z_j^{t+1} = \max\{0, y_j^t - x_j^t\}$, it can be shown that the results derived below still apply.

Each country produces under constant returns to scale, with average cost $c_j$. Each country, being risk neutral, maximizes the expected present value of its stream of rents over an infinite time horizon.

Under the stated assumptions, for each country $j$ there is a price $p_j^*$ and a supply $y_j^*$ such that (provided $z_j^1 \leqslant y_j^*$) there exists a dynamic equilibrium with the property that $p_j^t = p_j^*$ and $y_j^t = y_j^*$ for all $t = 1, 2, \ldots$ (This theorem is due to Kirman and Sobel [84]; see also Kawasaki, McMillan, and Zimmermann [72].)

Before a proof of this result is sketched, note that it says that, provided the initial level of stocks is not too high, there is an equilibrium in which each country charges a price which is constant over time and varies its outputs so as to maintain supply constant over time. From (7.1) and (7.2),

$$q_j^{t+1} = y_j^{t+1} - (y_j^t - x_j^t). \tag{7.3}$$

If $y_j^{t+1} = y_j^t = y_j^*$, (7.3) implies

$$q_j^{t+1} = x_j^t. \tag{7.4}$$

Equation (7.4) says that, in equilibrium, a fluctuation in demand is exactly matched by a change in output in the next period; there is no price response to a demand fluctuation.

Thus, given the crucial assumptions of identically, independently distributed demand fluctuations and constant-returns-to-scale production, it is in each country's interest to stabilize its price by using stockpiling and output changes to absorb the fluctuations in world demand. A proof of this result is sketched in the next section.

### 7.4* Constant-price equilibrium

The constant-price, constant-supply result can be understood as follows. The rent accruing to country $j$ in period $t$ is

$$\pi_j^t = p_j^t x_j^t - I_j(z_j^{t+1}) - q_j^t c_j^t. \tag{7.5}$$

This can be rewritten in terms of prices, demands, and supplies only, using (7.1) and (7.2):

$$\pi_j^t = p_j^t x_j^t - I_j(y_j^t - x_j^t) - [y_j^t - (y_j^{t-1} - x_j^{t-1})]c_j. \tag{7.6}$$

The country seeks to maximize $E[\sum_{t=1}^{\infty} \alpha_j^{t-1} \pi_j^t]$, where $\alpha_j$ is country $j$'s discount parameter. This game is a time-dependent game: that is, the payoff $\pi_j^t$ in any period depends on the value of past variables (namely $y_j^{t-1} - x_j^{t-1}$) as well as present variables. It can, however, be transformed into a simpler game, lacking time dependence. From (7.2), the term $[y_j^t - x_j^t]$ represents the amount of the commodity carried over in storage from period $t$ to period $t+1$. Such stocks are valuable to the country, because they reduce the amount that must be produced in period $t+1$. Given constant average cost of production, each unit of the commodity carried over in storage to period $t+1$ reduces period $t+1$ production cost by $c_j$; thus the value in period $t$ of one unit in storage is $\alpha_j c_j$. It follows that the discounted sum of the terms $\pi_j^t$ is the same as the discounted sum of the terms $\xi_j^t$, where $\xi_j^t$ is defined by

$$\xi_j^{t-1} = p_j^{t-1} x_j^{t-1} - I_j(y_j^{t-1} - x_j^{t-1}) - y_j^{t-1} c_j + (y_j^{t-1} - x_j^{t-1}) \alpha_j c_j. \tag{7.7}$$

(Roughly speaking, the $\pi_j^t$'s are transformed into the $\xi_j^t$'s by shifting $[y_j^{t-1} - x_j^{t-1}]$ back one term in the present-value summation.) This transformed game lacks time dependence. Taking expectations (remembering that $x_j^t$ is a random variable whose distribution depends on all the prices), write

$$E[\xi_j^t] = G_j(p^t, y^t), \tag{7.8}$$

where $p^t \equiv (p_1^t, p_2^t, \ldots, p_n^t)$ and $y^t = (y_1^t, y_2^t, \ldots, y_n^t)$. The problem has been reduced to finding an equilibrium for the nonstochastic repeated game in which country $j$ seeks to maximize the present value of the $G_j$'s. Each period's rent depends only on the prices, $p^t$, and supplies, $y^t$ (and not directly on the outputs, $q^t$, or the amounts carried over in storage, $z^t$). As was shown in Section 3.1, one of the equilibria of a time-independent repeated game is simply the repeated playing of a single-period-game equilibrium. Consider the static game in which country $j$'s objective is to choose $p_j$ and $y_j$ to maximize $G_j(p, y)$; denote the static Nash equilibrium of this game by $p^*, y^*$, where $p^* \equiv (p_1^*, p_2^*, \ldots, p_n^*)$ and $y^* \equiv (y_1^*, y_2^*, \ldots, y_n^*)$. Then an equilibrium of the repeated game, country $j$'s objective in which is to maximize $\sum_{t=1}^{\infty} \alpha_j^{t-1} G_j(p^t, y^t)$, involves country $j$ repeatedly playing $p_j^*, y_j^*$. Hence the constant-price, constant-supply result has been established.

### 7.5 Qualifications

By modelling the problem as a stochastic game, it has been shown that it is in a country's interest to stabilize the price of its exported commodity by adding to its stockpile of the commodity when world demand is low and drawing down its stockpile when world demand is high. This result is, however, sensitive to the model's assumptions. These assumptions are sufficient, not necessary, conditions for a constant-price policy to be an equilibrium; however, it is easy to show that relaxing some of the assumptions eliminates the desirability of holding prices constant.

One of the assumptions is that of constant-returns-to-scale production. Recall the transformation of the country's objective function from the sum of $\pi_j^t$'s to the sum of $\xi_j^t$'s. To carry out this transformation, it must be possible to assign a value to currently held stocks. The value of a unit of the commodity held in stock in reducing the next period's production costs is unambiguous only if unit production costs are constant. Hence constant returns to scale in production is crucial to the constant-price result.

The constant-price result requires that all demand fluctuations be purely transitory. If there is some autocorrelation in the demand fluctuations (as is likely to be the case if they are due to business cycles, for example), or if there is a permanent shift in demand, then may no longer be in the country's interest to keep its price constant.

There remain several open questions. First, the model was shown to be equivalent to a repeated game of the sort discussed in Section 3.1. One particular equilibrium, the analogue of the static Cournot-Nash equilibrium, was examined. As noted in Section 3.2, however, repeated games have other equilibria, including implicitly collusive outcomes. It remains an open question whether in this stochastic environment the countries could operate cartel strategies similar to those of Section 6.1. Second, it has been suggested that buyers' stockpiling increases the volatility of prices (Nurkse [119]; Brandt, *et al.* [20], p. 146); the possibility that buyers as well as producers hold stocks was not included in the above model. Third, game-theoretic ideas might help explain why attempts to establish international commodity agreements to stabilize prices have usually failed (Brandt, *et al.* [20], pp. 151–52). Fourth, the above model demonstrated sufficient conditions for the viability of what might be called perfect price stabilization: a less stringent notion of price stabilization might be defined as keeping prices within a certain range rather than holding prices constant (Brandt *et al.* [20], p. 152). The conditions under which there exists a decentralized oligopolistic equilibrium involving this more limited degree of price stabilization could be examined.

### 7.6 Bibliographic notes

On stochastic games, see Friedman [44], Chapter 10 and Owen [125], Chapter 5. The notion of myopic equilibrium, and a proof of its existence, is due to Sobel [148].

The importance, from the point of view of less-developed countries, of using stockpiling to stabilize world commodity prices was argued by Keynes [82], Nurkse [119] and Brandt *et al.* [20], Chapter 9. Commodity stockpiling in perfectly competitive world markets was modelled by Newbery and Stiglitz [116]. Newbery [115] modelled markets with a monopolistic producer.

The model of this section is essentially the same as that developed in another context by Kirman and Sobel [84]. The heuristic proof of Section 7.4 followed Kawasaki, McMillan, and Zimmermann [72].

## 8. INTERNATIONAL COMMON PROPERTY

It has long been believed that natural resources owned in common are overexploited. Each user knows that any of the resource left unextracted may be taken by some other user: thus there is an incentive to extract too quickly.

A resource may be common property because it is physically impossible to assign ownership of it to an individual or a country (for example, since fish can move freely, property rights in ocean fish are infeasible); alternatively, a resource may be common property simply because property rights to it have not yet been assigned (attaching property rights to manganese nodules lying on the seabed outside national jurisdictions would require some form of international agreement).

The world's largest common properties are the oceans. In an attempt to resolve or forestall disputes between countries over matters such as fishing rights and seabed mining, the United

Nations General Assembly in 1970 proposed the creation of an international sea law code. For a decade, the Law of the Sea Conference discussed the complicated questions of how the ownership of ocean resources should be assigned to nations. A treaty was adopted at the end of 1982, but several nations refused to sign it.

While many of the sources of disagreement in the Law of the Sea Conference were questions of equity, one of the main points of agreement was one of efficiency: this was the belief that, in the absence of well-defined property rights, the world's ocean resources would be extracted too quickly. This section develops a model of the extraction of an international common-property exhaustible resource. The value of the model is that it exposes the implicit assumptions underlying the conventional assertion that common-property resources are depleted too quickly. Since the conventional assertion refers to the speed of extraction of the resource, an adequate model must be dynamic: the common-property problem is modelled here as a differential game.

### 8.1 Pareto optimal extraction

Consider the extraction of an exhaustible natural resource. For the sake of tractability, make two assumptions which are frequently used in exhaustible-resource models: first, extraction is costless; and second, the demand for the resource is given by an isoelastic demand function, with the price elasticity of demand exceeding one in absolute value. Denote by $r(t)$ the amount extracted and sold at time $t$, by $s(t)$ the size of the stock of the resource at time $t$, and by $p(r)$, $p' < 0$, the instantaneous demand function. The elasticity $\eta \equiv -p/rp' \geqslant 1$ is constant.

The Pareto optimal rate of extraction is given by the Hotelling rule: the price of the resource should rise at the rate of interest, $\alpha$. That is, using $\hat{z}$ to denote $\dot{z}/z$, the Hotelling rule says

$$\hat{p} = \alpha. \tag{8.1}$$

Given constant price elasticity of demand, and given that $\hat{s} = -r/s$, (8.1) implies

$$\hat{s} = -\eta\alpha. \tag{8.2}$$

There are two possible distortions on the supply side of a natural-resource market. First, an extractor may have some market power in selling the resource. Second, common-property aspects might cause the rate of extraction to be inappropriate. In this model, if the resource is extracted monopolistically, neither of these distortions arises. Because the elasticity of demand is assumed to be constant, there is no distortion due to monopoly power when the resource is sold. And with only one extractor, common-property problems clearly do not arise. Thus the Pareto optimal rate of extraction (8.1) or (8.2) would be achieved if the resource were controlled by a single decision-maker.

Suppose now that two countries have access to the common-property exhaustible resource. Each country chooses an extraction strategy to maximize the discounted value of its stream of future rents, given the other country's extraction strategy. This is an example of a differential game.

## 8.2 Differential games

The dynamic game discussed in Section 3.1 had one particular simplifying characteristic: the structure of the game did not change over time. Consider now a type of game in which change is inherent. Suppose there is some state variable which defines the current state of the game; and the value of this state variable depends on each of the agents' actions in the past. If such a problem is modelled in continuous time, it is a differential game.

Two alternative types of strategy are commonly assumed in differential games. An *open-loop strategy* specifies each agent's action as a function of time alone. In effect, at the initial date, each agent commits himself to a trajectory of actions for all future dates. An *open-loop Nash equilibrium* is a set of open-loop strategies, one for each agent, such that the path to which each agent commits himself is a best response to the paths to which the other agents have committed themselves.

A *feedback strategy* specifies each agent's action as a function of the state variable as well as time, for all possible values of the state variable. A feedback strategy does not require an agent to commit himself at the initial date to a particular trajectory of actions; instead, the agent commits himself at the initial date to a decision rule. At each point in time, each agent can observe the current value of the state variable, and choose his best action accordingly. A *feedback Nash equilibrium* is a set of feedback strategies (decision rules), one for each agent, such that the decision rule which each agent chooses is a best reply to the decision rules which all the other agents have chosen.

The open-loop equilibrium usually yields an allocation of resources different to the feedback equilibrium.

Along an open-loop equilibrium path, at any point in time each agent is doing the best he can given the other agents' strategies: this simply says that the time path is, indeed, an equilibrium. Feedback strategies satisfy a stronger criterion: at any point in time, each agent, by following his feedback strategy, is optimizing given the others' strategies, regardless of whether or not the system is on the equilibrium path. Thus the feedback equilibrium is a subgame perfect equilibrium.

Notice that neither of these solution concepts incorporate the kind of behavior discussed in Section 3.2 where agents respond to the past actions of the other agents. Recall that it is these reactions (noncooperative "threats" and "punishments") which enable cooperative-like behavior to occur in dynamic noncooperative games. How to model such dynamic strategies in differential games remains an open question.

### 8.3 Equilibria of the common-property game

What are the equilibria for the differential game of extraction of a common-property resource? In Sections 8.4 and 8.5 the following two results are proven (following McMillan and Sinn [109]). In the open-loop equilibrium, the resource is extracted at the Pareto optimal rate. In the feedback equilibrium, the resource is exhausted immediately.

Both results are striking. In the open-loop equilibrium, despite the fact that the two countries are extracting from a common pool, they do not extract too quickly. Evidently, the traditional view that common-property results in overextraction is contradicted by the open-loop equilibrium.

The feedback equilibrium, on the other hand, dramatically vindicates the conventional wisdom that common-property resources are overexploited: there is the maximal amount of overextraction. Note that the feedback equilibrium with only two countries extracting the resource results in the same outcome as would free entry of extractors.

### 8.4* Open-loop equilibrium

Open-loop strategies require extraction rates to be functions of time alone. Suppose country 2 is using an open-loop strategy so that its entire extraction plan is given by the function $r_2(t)$. What is country 1's best reply $r_1(t)$ to this strategy?

Given the initial size of the stock $s(0) > 0$, country 1's objective is to choose an extraction plan $r_1(t)$, with $r_1(t) \geqslant 0$, to maximize

$$\int_0^\infty e^{-\alpha t} r_1(t) p(r_1(t) + r_2(t))\, dt. \tag{8.3}$$

The Hamiltonian is

$$e^{-\alpha t}[r_1 p(r_1 + r_2) - (r_1 + r_2)\lambda_1], \tag{8.4}$$

where $\lambda_1$ is the shadow variable. Necessary conditions (assuming an interior solution) are

$$\lambda_1 = p(r_1 + r_2)\left(1 - \frac{r_1}{\eta(r_1 + r_2)}\right), \tag{8.5}$$

and

$$\dot{\lambda}_1 - \alpha\lambda_1 = 0. \tag{8.6}$$

The transversality condition is

$$\lim_{t \to \infty} e^{-\alpha t}\lambda_1(t)s(t) = 0. \tag{8.7}$$

Because $\hat{\lambda}_1 = \alpha$ (from (8.6)), the transversality condition reduces to

$$\lim_{t \to \infty} s(t) = 0. \tag{8.8}$$

Conditions (8.5), (8.6), and (8.8) determine country 1's best extraction plan given country 2's extraction plan.

Country 2 similarly chooses its extraction plan given country 1's extraction plan. At an equilibrium, country 2's best extraction plan must be the same as the extraction path for country 2 which country 1 assumed in making its own optimizing decision; and vice versa. Each country's equilibrium extraction plan is a best reply to the other country's equilibrium extraction plan.

Consider only symmetric equilibria. Thus, in equilibrium, $(r_1 + r_2)/r_1 = 2$, so that $(1 - r_1/\eta(r_1 + r_2))$ is constant. Hence, in equilibrium, (8.5) implies

$$\dot{\lambda}_1 = p'\dot{r}\left(1 - \frac{1}{2\eta}\right), \tag{8.9}$$

where $r = r_1 + r_2$. From (8.5) and (8.9)

$$\hat{\lambda}_1 = -\frac{\hat{r}}{\eta}. \tag{8.10}$$

From (8.6) and (8.10),

$$\hat{r} = -\alpha\eta. \tag{8.11}$$

The transversality condition (8.10) implies that

$$s(t) = \int_t^\infty r(\tau)\, d\tau$$

$$= \frac{r(t)}{\alpha\eta}, \tag{8.12}$$

the last equation following from (8.11). Thus, from (8.11) and (8.12)

$$\hat{s} = -\frac{r}{s} = -\alpha\eta. \tag{8.13}$$

Comparing (8.2) and (8.13) gives the result that, in the open-loop equilibrium, the resource is extracted at the Pareto optimal rate.

### 8.5* Feedback equilibrium

Feedback strategies depend upon the size of the stock of the resource as well as time. Thus a feedback strategy specifies that the rate of extraction by country 2 at time $t$ is

$$r_2(t) = \Psi_2(t, s(t)), \tag{8.14}$$

where $\Psi_2$ depends upon s nontrivially. For simplicity, attention will be restricted to linear feedback strategies:

$$r_2(t) = a_2(t) + b_2 s(t), \tag{8.15}$$

where $b_2$ is constant, $0 < b_2 < \infty$, and $a_2(t) + b_2 s(t) \geqslant 0$. (Note, for comparison, that if $b_2 = 0$, this reduces to the open-loop case of the last section.) The term $a_2(t)$ in the strategy (8.15) indicates that, in part, country 2's extraction rate is autonomous. The term $b_2 s(t)$ indicates that an increase in the size of the resource stock will feed back into country 2's actions, causing country 2 to increase its extraction rate. In particular, if country 1 leaves one extra unit of the resource unextracted, country 2 will immediately extract a fraction $b_2$ of it.

As in the last section, consider country 1's problem of finding an extraction plan $r_1(t)$ which is a best reply to country 2's extraction plan (8.15). Country 1's objective is to choose $r_1(t) \geqslant 0$ to maximize

$$\int_0^\infty e^{-\alpha t} r_1(t) p(r_1(t) + a_2(t) + b_2 s(t))\, dt. \tag{8.16}$$

The Hamiltonian is

$$e^{-\alpha t}[p(r_1 + a_2 + b_2 s) r_1 - \mu_1(r_1 + a_2 + b_2 s)], \tag{8.17}$$

where $\mu_1$ is the shadow variable. Necessary conditions, assuming an interior solution, are:

$$\mu_1 = p(r_1 + a_2 + b_2 s)\left[1 - \frac{r_1}{\eta(r_1 + a_2 + b_2 s)}\right], \tag{8.18}$$

and

$$\dot{\mu}_1 - \alpha\mu_1 = -b_2(p' r_1 - \mu_1). \tag{8.19}$$

The transversality condition is

$$\lim_{t \to \infty} e^{-\alpha t} \mu_1(t) s(t) = 0 \tag{8.20}$$

Using (8.18) in (8.19), note that

$$\hat{\mu}_1 = \alpha + b_2\left[1 + \frac{1}{\dfrac{\eta(r_1 + a_2 + b_2 s)}{r_1} - 1}\right]. \tag{8.21}$$

Given $\eta \geqslant 1$, (8.21) implies $\hat{\mu}_1 \geqslant r$, which in turn implies that (8.20) reduces to

$$\lim_{t \to \infty} s(t) = 0. \tag{8.22}$$

Country 1's best reply to country 2's decision rule (8.15) is given by (8.18), (8.19), and (8.22). Country 2 solves a similar problem, given country 1's decision rule $r_1(t) = a_1(t) + b_1 s(t)$. At a feedback equilibrium, each country's decision rule is a best reply to the other country's decision rule.

Attention will be confined to symmetric equilibria of the form (8.15): thus $a_1(t) = a_2(t) = a(t)$; $b_1 = b_2 = b$. In equilibrium, $r/r_1 = 2$, so that $(1 - 1/(\eta r/r_1))$ is constant. Thus, in equilibrium, (8.18) and (8.21) imply

$$\hat{\mu}_1 = -\frac{\hat{r}}{\eta}. \tag{8.23}$$

Thus

$$\hat{r} = -\eta\left[\alpha + b\left(1 + \frac{1}{2\eta - 1}\right)\right]. \tag{8.24}$$

Equation (8.24) implies that

$$\int_t^\infty r(\tau)d\tau = \frac{r(t)}{\eta\left[\alpha + b\left(1 + \dfrac{1}{2\eta - 1}\right)\right]}. \tag{8.25}$$

Since, from (8.22), $s(t) = \int_t^\infty r(\tau)\, d\tau$, (8.25) implies

$$r(t) = \eta\left[\alpha + b\left(1 + \frac{1}{2\eta - 1}\right)\right]s(t). \tag{8.26}$$

Since the equilibrium is assumed to be symmetric, each country's extraction rate is one-half of the total extraction rate; thus in particular

$$r_2(t) = \frac{1}{2}\eta\left[\alpha + b\left(1 + \frac{1}{2\eta - 1}\right)\right]s(t). \tag{8.27}$$

At a feedback equilibrium, the function representing any country's extraction plan must be the same function as that country's best reply to the other's extraction plan. Thus (8.15) and (8.27) must be the same function; in particular, the coefficients of $s(t)$ must be equal. Thus, at a feedback equilibrium

$$b = \frac{1}{2}\eta\alpha + \frac{1}{2}\eta\left(1 + \frac{1}{2\eta - 1}\right)b. \tag{8.28}$$

Elementary manipulations show that $\eta \geqslant 1$ implies

$$\frac{1}{2}\eta\left(1 + \frac{1}{2\eta - 1}\right) \geqslant 1. \tag{8.29}$$

Thus, since $\eta\alpha/2 > 0$, there is no finite value of b which satisfies (8.28) with the right-hand side of (8.28) non-negative; that is (from (8.26)), with non-negative rates of extraction.

What has been shown is that there exists no feedback equilibrium involving linear strategies with finite extraction rates.

Suppose the resource could be extracted instantaneously. This would correspond to the limiting case of countries using strategies of the form (8.15) with $b_1$ and $b_2$ approaching infinity. The degenerate case of instantaneous extraction satisfies the definition of a feedback equilibrium. (In the limit, equation (8.28) holds as $b$ approaches infinity.) Thus the preceding analysis has established that the only feedback equilibrium in linear strategies involves instantaneous extraction of the whole stock of the resource. (In fact, the analysis of Reiganum and Stokey [133] shows that the restriction to linear strategies is redundant: the instantaneous-extraction result holds for general functional forms.)

### 8.6 Comparison of the equilibria

The two concepts of open-loop and feedback equilibria yield sharply different predictions about the rate of extraction of a common-property resource: in the open-loop equilibrium, there is no overextraction; but in the feedback equilibrium there is the maximal amount of overextraction. What is the reason for this difference? In the light of these results, what is the status of the traditional view that common-property resources are depleted too quickly?

Pareto optimal extraction in the open-loop equilibrium occurs because each country behaves as if it had well-defined property rights. It believes that the other country maintains a given extraction path independently of its own actions. There is a given quantity of the resource available for it to extract: there is no need to speed up its extraction process in order to preclude extraction by the other country. In effect, open-loop strategies implicitly exclude the common-property interactions which underlie the traditional assertion about overexploitation.

With feedback strategies, there is a direct interaction between the two countries' decisions. Feedback strategies mean that a change in the size of the resource stock will cause a change in each country's extraction rate. If country 2 is using a strategy of the form (8.15), then country 1 knows that, of every unit of the resource it leaves unextracted, a fraction $b_2$ will immediately be extracted by country 2; this amount of the resource is therefore lost to country 1. Thus country 1's valuation of the unextracted resource is too low: it extracts the resources too quickly; and similarly for country 2. Since, as was shown, there are no mutually consistent feedback strategies with finite values of $b_2$, country 1 knows that country 2, if it is using a feedback strategy, will immediately extract all of the resource which country 1 leaves unextracted.

The Pareto optimality of the open-loop equilibrium suggests a direction of approach for legislative attempts to overcome common-property problems, such as the Law of the Sea Conference. Remedies adopted within national economies for the distortions which arise when several firms extract from a common pool of oil take the form, for example, of prorationing (fixing a maximum permissible rate of extraction by individual firms) or of compulsory or voluntary unitization (operating the whole pool under a single decisionmaker and then distributing the profits among the individual firms: that is, collusion among the firms). In effect, these policies force the equilibrium to be open loop.

The analysis assumed that the initial size of the stock of the resource, $s(0)$, was known to all countries. If, on the contrary, the location and size of the resource stock were unknown, there would have to be initial exploration. With inadequate property rights, there would typically be too little exploration: a country incurring exploration costs would not gain the full benefit from its search if other countries were able to extract the resource after discovery. (The information is a public good: compare with Section 10.1.) Notice that this distortion works in the opposite direction to the common-property distortion.

### 8.7 Bibliographic notes

On differential games, and the distinction between open-loop and feedback equilibria, see Starr and Ho [150, 151].

The economic issues involved in the Law of the Sea negotiations were discussed by Sweeney, Tollison, and Willett [154] and Wijkman [163]. Raiffa [132], Chapter 18 described the process of negotiating the treaty.

International common-property problems were analyzed in static models by Anderson [4] and Markusen [100, 101]. The analysis of Sections 8.4 and 8.5 followed McMillan and Sinn [109]. Open-loop solutions for the dynamic common-property problem were analyzed by Bolle [15], Chiarella, *et al.* [24], and Kemp and Long [79]. Feedback solutions for such a game were derived by Clark [27], Eswaren and Lewis [43], Levhari and Mirman [94], and Reinganum and Stokey [133]. Sinn [147] provided an alternative solution, under which extractor believes that its rivals will extract so as to maintain a given time-path of the stock. Munro [111] presented a cooperative-game dynamic model of an international fishery. The informational externalities from exploration were modelled by Quyen [130].

## 9. CUSTOMS UNION AND THE CORE

A customs union is formed when two or more nations eliminate restrictions on trade between themselves and set a common tariff on trade with the rest of the world. The most prominent example of a customs union is the European Economic Community; other customs unions, existing or defunct, include the Andean Pact, the East African Common Market, and the New Zealand-Australia Free Trade Agreement.

A basic question, much discussed but not resolved in the early writings by economists on economic integration, asks whether a customs union should be regarded as a move in the direction of, or away from, worldwide free trade. It is a move towards free trade in that barriers to trade among member countries are removed. It is a move away from free trade in that restrictions on trade outside the union are solidified; the theory of second-best warns that, even though the elimination of all tariffs would increase total world welfare, piecemeal tariff reductions may lower world welfare.

To be successful, a customs union must solve some internal problems. Members must agree on the common levels of tariffs and quotas to be imposed on the rest of the world. These tariffs and quotas have distributional effects both within member countries and across member countries. (One of the sources of conflict among member countries in the EEC is the extent to which agriculture should be protected from international competition.) In order that all members share the gains accruing from union, there may need to be transfers among union members. (Deciding the size of these transfers is an annual problem for the EEC.) Internal pressures of these kinds may accumulate enough to destroy the customs union (as in the case of the East African Common Market).

This section uses the concept of the core to analyze the formation of customs unions. The model sheds some light on the question of whether economic integration represents a move towards international free trade; and it shows the role of the common external tariff and the transfers across member states in ensuring that all members gain from the union.

### 9.1 The core and competitive equilibrium

Before discussing the theory of customs unions, it is necessary to recall some fundamental results from modern general-equilibrium theory.

Consider a model in which each individual consumer and producer is a price-taker: the trade of each is too small a fraction of total trade to have any influence on the market price. At a competitive equilibrium, each producer maximizes profits, each consumer maximizes utility given initial endowments, and supply equals demand for each commodity. Under the well-known Arrow-Debreu assumptions (ruling out increasing returns to scale, for example), the first theorem of welfare economics says that any competitive equilibrium in Pareto optimal; and the second theorem of welfare economics says that, for any Pareto optimum, there is a price vector under which that Pareto optimum can be attained as a competitive equilibrium (Debreu [33], Chapter 6).

Consider now in more detail the concept of the core introduced in Section 2.2. Denote the set of all agents by $\mathcal{N}$, and let a coalition, $S$, be any subset of $\mathcal{N}$. The definition of the core given in Section 2.2 assumed transferable utility; it is, however, easily extended to the nontransferable-utility case. The question to be asked is: can some coalition of agents, by their own actions, make themselves better off? An allocation of resources, $x$, can be *blocked* by coalition $S$ (or *improved upon* by coalition $S$) if there exists another allocation $x'$, attainable through the actions of the members of $S$ alone, such that

$$u^i(x') > u^i(x) \text{ for all } i \in S, \text{ and for all feasible } x'_j, j \notin S. \quad (9.1)$$

Then the core is the set of allocations that are feasible and not blocked by any coalition. That is, an outcome is in the core if no coalition can find an alternative set of strategies which make every member of the coalition better off regardless of what strategies the nonmembers adopt.

Notice that the definition of Pareto optimality can be rephrased: a Pareto optimum is an allocation of resources which is not blocked by the coalition consisting of all agents. Thus any allocation in the core is Pareto optimal.

For an economy with a finite number of agents, the set of competitive equilibria is a strict subset of the core: there are core allocations which are not competitive equilibria. However, as the number of agents in the economy increases, the core shrinks. In the limit, with infinitely many agents, it can be shown that the core coincides with the set of competitive equilibria (Edgeworth [42]; Arrow and Hahn [6], Chapter 8).

### 9.2 The core and free trade

Consider a model with any number of countries, but within each country there are many price-taking consumers and producers. Assume that, at the level of the individual producer and individual consumer, the usual assumptions of the Arrow-Debreu model are satisfied (Arrow and Hahn [6], Debreu [33]).

Applying the result on the equivalence of the core and the set of competitive equilibria, the core of this international economy consists of the set of free-trade allocations. If it were possible freely to form coalitions of individual consumers, then any non-free-trade allocation could be improved upon by some coalition. In the context of the international economy, however, coalitions cannot be arbitrarily formed. To some extent the coalition structure is fixed exogenously. All the citizens of a particular country must belong to the same coalition; coalitions must consist of countries, or of groups of countries. A coalition of countries is a customs union.

In this model, with restricted opportunities for coalition formation, is there any reason to suppose that core is the same as with unrestricted coalition possibilities, namely free trade? An answer is provided by a proposition due to Kemp and Wan [81].

Consider a competitive world trading equilibrium, with arbitrary initial tariffs. Now let any subset of the countries form a customs union. Then there exists a common external tariff vector and a system of lump-sum transfers among the members of the union such that, at the new world competitive equilibrium, each individual citizen (of both member and nonmember countries) is at least as well off as before union.

This proposition is proved as follows (directly following Kemp and Wan). Starting with a tariff-ridden world competitive equilibrium, consider some coalition of countries. Imagine a hypothetical economy composed of the member countries except that its net endowment is the sum of members' endowments plus the pre-union net excess supply of the rest of the world. By the first and second theorems of welfare economics, this hypothetical economy has a Pareto optimum which can be supported by some price vector; at this Pareto optimal equilibrium there will be no restrictions on trade between member countries. Either (a) the pre-union equilibrium of the member countries is a Pareto optimal equilibrium of the hypothetical economy, or (b) it is not. In case (a), the union does not change the world equilibrium. In case (b) an equilibrium for the hypothetical economy which is a Pareto improvement over the members' pre-union situation can be attained by means of lump-sum transfers among individual consumers in the hypothetical economy. Now set the vector of common external tariffs to be the difference between the vector of pre-union world prices and the vector of prices in the hypothetical economy. Because of its tariff policy, this union does not change world prices, and therefore net excess supply from the rest of the world remains unchanged and utility levels in the rest of the world are unaltered. The allocation within the hypothetical economy can be attained within the union. This completes the proof. A customs union formed in this way makes no individual worse off and usually makes some better off.

The Kemp-Wan construction almost, but not quite, proves that any tariff-ridden world equilibrium can be blocked by some customs union. To establish this, it remains to show that union members can be made strictly better off, rather than just no worse off. Suppose the hypothetical economy has positive internal tariffs before union. It has zero tariffs after union. Removing this distortion creates a net welfare gain within the hypothetical economy; lump-sum transfers can be used to ensure all individuals in member countries gain.

Thus, a coalition of countries can be found which improves upon any tariff-ridden world equilibrium.

### 9.3 Customs union and free trade

This model provides some support for the view of a customs union as a move towards free trade. The result applies to any set of initial conditions. After one customs union has formed, incentives remain either for the customs union to accept new members, or for further customs unions to form. These incentives for expanding customs unions persist until the core, that is worldwide free trade, is reached (Kemp and Wan [81], p. 96).

Two qualifications to this conclusion must be made, perhaps in part explaining why in the world, as opposed to the model, free trade is not attained.

First, the analysis demonstrated the possibility that countries could set their tariffs so as to make a customs union beneficial: tariff levels were set so that the union's trade with the rest of the world remains at its pre-union quantity. It is not, however, necessarily in each country's interest to follow a tariff policy of this sort. If a customs union were large relative to the rest of the world (as eventually it would have to become in the envisaged process of successively larger customs unions), the union would have some power to influence world prices (compare with Section 4.1). While setting tariffs so as to exploit this market power would make member-countries better off, it might halt the convergence to free trade.

Second, for the gains from union to be enjoyed by all individuals in member countries, it may be necessary that intra-union transfers take place. For such transfers to be achieved, member countries must give up some of their national sovereignty. Agreements made across member-country borders must be enforceable; the intra-union game must be a cooperative game. If agreements on transfers are not enforceable so that the intra-union game is a noncooperative game, the suboptimality of static noncooperative equilibrium may be relevant (see Section 2.3). It is not enough that all member countries perceive the mutual gains to be had from a system of transfers; there must be some mechanism for deciding the level of transfers and for carrying out the transfers.

### 9.4 Bibliographic notes

The analysis of this section followed Kemp and Wan [81]; for similar analyses, see Dixit and Norman [37], pp. 191–194, Grinols [54], and Ohyama [120]. On the sources of welfare gains from economic integration, see Kemp [74], McMillan and McCann [108], and the references therein. Segal [143] and Gately [48] used respectively the Shapley value and the core to model the distribution of the gains within a customs union. Mantel and Martirena-Mantel [98] compared a cooperative outcome (a customs union) with a noncooperative outcome (a tariff Nash equilibrium of the sort modelled in Section 4.3 above). Ichiishi and Riezman [66] modelled customs-union formation with countries setting tariffs in retaliatory fashion.

## 10. INTERNATIONAL TRANSFERS

An old question in international economics (dating back at least as far as Keynes's discussion of the post-World-War I German reparations) is about the effects of international transfers. Transfers of money between countries are common: they take the form not only of war reparations but also of foreign aid, or of remittances by emigrant workers, or of international borrowing and lending. These international transfers can have surprising effects.

In one simple but plausible model (to be presented in Section 10.1), international transfers have no effect whatsoever on any country's consumption or production. This striking conclusion relies on the presence of some international public good, jointly produced and consumed by all countries' citizens.

In another model (summarized in Section 10.2), transfers do change production and consumption; indeed, it is possible that a transfer could result in the donor country's consumption increasing and the recipient country's consumption decreasing. This perverse possibility relies on the transfer having a large effect on the world terms of trade.

A third class of models (briefly noted in Section 10.3) examines international borrowing and lending, seeking explanations of such phenomena as defaults by borrowers and credit limits imposed by lenders.

### 10.1 International transfers and international public goods

The theory of public goods can be used to model some kinds of interactions among nations. One example of an international public good is the control of international pollution, such as the acid rain falling on both the United States and Canada or the pollution of the Rhine generated by upstream countries and affecting downstream countries. Another example is the mutual provision of defense through alliances such as the North Atlantic Treaty Organization. Yet another example is unpatented basic scientific research.

Suppose $n$ countries share in the benefits of an international public good. The public good is produced through the contributions of the individual nations. Nation $i$ contributes an amount $g_i$ towards the provision of the public good, so that the total supply of the public good is $G = \sum_{i=1}^{n} g_i$. In addition, the citizens of nation $i$ consume a domestically-produced private good in amount $x_i$. Given the world price, $p$, of the private good and country $i$'s national income at world prices, $m_i$, country $i$'s world budget constraint is

$$px_i + g_i = m_i,\ i = 1, \dots, n. \tag{10.1}$$

Each nation has a community preference function $u^i(x_i, G)$. Using (10.1), each nation maximizes utility given the others' contributions to public-good provision by contributing an amount $g_i$ which solves (assuming an interior solution)

$$\frac{u_2^i}{u_1^i} = \frac{1}{p}. \tag{10.2}$$

The static Nash equilibrium of public-goods provision is found by solving simultaneously n equations of the form (10.2).

Compare the Nash equilibrium with the Pareto optimal provision of the public good; this is found by solving the Samuelson condition:

$$\sum_{j=1}^{n} \frac{u_2^j}{u_1^j} = \frac{1}{p}. \tag{10.3}$$

Comparing (10.2) and (10.3), and using the strict concavity of the utility function, it can be shown that too small an amount of the public good is supplied at the static Nash equilibrium. A quick argument demonstrating this is as follows. Comparing (10.2) and (10.3) shows that $u_2^i/u_1^i$ must be larger at the Nash equilibrium than at a Pareto optimum; thus $u_2^i$ is larger and $u_1^i$ is smaller than optimal at the Nash equilibrium; thus the public good is undersupplied at the Nash equilibrium. Each country tries to free-ride on the benefits of the public good supplied by the other countries, so too little public good is supplied. (However, as discussed earlier in other contexts, if this game can appropriately be modelled as a dynamic game, the repetition of plays may result in a Pareto optimal dynamic equilibrium being reachable.)

Suppose now that transfers among the countries are possible. Continue to assume an interior solution (10.2), so that all countries contribute something towards public-good supply.

Denote by an "N" superscript the Nash-equilibrium value of a variable. It can be shown that this game has a unique Nash equilibrium, found by solving the $n$ simultaneous equations (10.2). Solving (10.2),

$$x_i^N = x_i^N(p, G^N). \tag{10.4}$$

From (10.1),

$$M \equiv \sum_{j=1}^{n} m_j = p \sum_{j=1}^{n} x_j^N(p, G^N) + G^N \tag{10.5}$$

(10.5) can be solved for $G^N$:

$$G^N = G^N(p, M). \tag{10.6}$$

Finally, substitute (10.6) in (10.4):

$$x_i^N = x_i^N(p, M). \tag{10.7}$$

Equations (10.6) and (10.7) contain the result. The quantity of the public good consumed within each country (which by definition of a public good is the same as the total amount available) and the quantity of the private good consumed within each country depend only on aggregate income, not on each country's income. Hence a transfer of income from one country to another does not change any country's consumption. A transfer results in the donor offering less towards public-good supply and the recipient offering more; these two effects exactly cancel each other.

The assumption of an interior solution was crucial to this result. If a country were sufficiently poor, it would contribute nothing towards public-good provision (equation (10.2) would be replaced by an inequality). Then a change in the distribution of income across countries might induce a country formerly not contributing to public-good provision to begin contributing, or vice versa. If this were to happen, the quantity of public good supplied and the quantities of the private good consumed would change following an international transfer.

### 10.2 International transfers and the terms of trade

Suppose, reverting to a more standard trade model, either that there are no public goods, or that goods are public only within any nation's borders.

Can foreign aid make the donor country better off and the recipient country worse off? Could a gift from one country to another make both better off while harming a third country? Examples can be constructed giving affirmative answers to both of these questions.

There are two consequences of a gift from one country to another. The direct effect is obvious: the gift lowers the disposable income and therefore the consumption of the donor, and raises the consumption of the recipient. The indirect effect of these changes in consumption is, if the countries' imports and exports are large relative to the volume of world trade, to change the world terms of trade. Examples have been constructed in which the indirect effect outweighs the direct effect; that is, it is possible that a transfer can so much change the world terms of trade as to generate the apparently perverse welfare effects of the donor gaining and the recipient losing.

In the existing theoretical analysis of this question, there are only two or three countries. These countries have the ability, by changing their consumption, to change world prices: this is the very source of the apparently paradoxical welfare gains and losses. However, the countries are, in these analyses, assumed to act like small countries, as though their terms of trade were exogenously fixed. Thus the perverse welfare effects have only been shown to be possible if the countries act irrationally. It remains an open question whether similar results can be obtained in a model involving rational behavior. In such an analysis the countries would recognize their mutual interdependencies: the analysis would necessarily be game theoretic.

### 10.3 International loans

In response to topical concern about the amount of international borrowing and the threats of default by some less-developed countries, there have been developed several theoretical analyses of the international loan market. There is an important distinction between borrowing internationally and borrowing within a national economy. National bankruptcy laws require creditors to be compensated to the extent possible; bankruptcy laws prevent an individual from refusing to pay his debts while retaining control over his assets. In contrast, in the noncooperative-game context of the international economy, there is nothing equivalent to domestic bankruptcy laws to prevent a country from refusing to pay its external debts. Loans will be repaid, and therefore an international loan market can function, only if it is in a country's self-interest to repay its loans. The source of the incentive to repay debts is the dynamic-game nature of the situation: failure to repay reduces the likelihood of being offered a loan in the future.

The existing analyses of international borrowing contain combinations of some of the following elements (tractability rules out putting all of these features into a single model). The lender is modelled as a Stackelberg leader and the borrower as a Stackelberg follower. Uncertainty is a crucial element of these models: in the absence of uncertainty, a rational lender would simply lend no more that it would be in the borrower's interest to repay, given the borrower's known future income. Not only is there uncertainty, but also there is asymmetric information. A lender is not as well informed as a borrower: for example, a lender may not be able to observe the riskiness of a borrower's investment projects. As a result of the asymmetry of information, complete contingent contracts are infeasible: the provisions of a contract cannot be enforced if they are made contingent on events which one party does not observe. In consequence, there is a moral hazard problem: a borrower has an incentive to choose riskier projects than would be chosen were contingent contracts available (riskier projects, that is, than would be undertaken at a

Pareto optimum). A second moral hazard problem arises because the borrower determines his own rate of consumption, which affects the probability of default. As a result of the moral hazard and the incompleteness of contracts, the ability to choose price (interest rate) alone does not give the lender sufficient control over the relationship. A second instrument used by the lender is a credit limit: the borrower is allowed to borrow as much as he wants up to some limit specified by the lender. This gives the lender partial control over the borrower's consumption rate and therefore default risk. The lender can provide two types of service to the borrower: the smoothing over time of the borrower's consumption stream, and the absorbing of part of the risk associated with the borrower's future consumption. The borrower is presumed to default on a loan whenever defaulting is less costly than not defaulting. The cost to the borrower of defaulting comes from reputational effects: a default lowers the probability of being able to negotiate a loan in the future. Typically in equilibrium the risk-sharing and consumption-smoothing benefits are achieved at the expense of a positive probability of default.

### 10.4 Bibliographic notes

The results on international transfers with public goods are due to Bergstrom, Blume, and Varian [10], Kemp [76], and Warr [161]. A Nash-equilibrium model of the provision of defense as an international public good was developed by Olson and Zeckhauser [121].

A large amount has been written on when a gift can make the recipient country or a third country worse off: for a sample, see Bhagwati, Brecher, and Hatta [13], Chichilnisky [25], Jones [70], Kemp and Kojima [77], and Leonard and Manning [93].

For models of international borrowing strategies, see Eaton and Gersovitz [39], Hellwig [61], and Kletzer [85]; for a survey of this literature, see Crawford [29].

## 11. COUNTRIES AS COALITIONS

This section identifies a fundamental link between the theory of games and the theory of international trade. A basic simplifying assumption often used in game theory is that utility is transferable. A basic simplifying assumption often used in international economics is that consumers' choices can be represented by a community preference function. In this section it is shown that the existence of a community preference function is equivalent to the transferability of utility within a small, open economy.

### 11.1 Transferable utility

The term "transferable utility" is a misnomer. Utility cannot, of course, be transferred from one individual to another; only commodities can be transferred. Suppose, however, there exists a commodity which can freely be transferred among the agents; this commodity can be thought of as being money. Examples of such transfers are bribes and side-payments. Roughly speaking, utility is transferable if utility functions are such that any transfer of money from one agent to another leaves the sum of agents' utilities unchanged. This is not, however, an adequate definition, because in most of economics utility is an ordinal concept: a utility function is defined only up to a monotonic increasing transformation; the choice of origin and units for a utility scale are arbitrary. Thus the sum of agents' utilities is not a well-defined concept. The complete definition states that utility is transferable if there exist monotonic increasing transformations of the agents' utility functions such that, when money is transferred among the agents, the sum of agents' utilities is constant.

Consider a game in which a fixed amount of total money income is to be divided among the agents. Then utility is transferable if and only if there can be found a set of utility scales such that the utility-possibility frontier is a straight line.

The analysis of cooperative games is simplified by assuming the transferability of utility. In cooperative games with more than two agents, there arises the possibility of coalitions. Transferable utility allows the separation of the analysis of coalition behavior into two parts. With transferable utility, the objective of the coalition is simply to maximize total coalition utility; the question of the division of the spoils among coalition members can then be decided as a separate matter. The coalition's best action is independent of the distribution of income within the coalition.

### 11.2 Community preference functions

Since the focus of international-trade theory is international, it is often convenient to abstract from the transactions which occur within a nation. One way of achieving this simplification is to assume that individual consumers' preferences can be aggregated to form a community preference relation; the nation as a whole behaves as if it were a single consumer.

The usefulness of some aggregate notion of community welfare is illustrated by one of the most important results of trade theory: the gains-from-trade theorem. The argument is illustrated by Figure 11.1. A nation produces two goods in amounts X and Y. Its production possibility frontier is AFDB. In the absence of international trade it can consume only what it produces; it must consume somewhere in the region OAFDB. Suppose now it freely trades internationally. The nation's trade is assumed to be a small enough fraction of world trade that it cannot affect world prices: the nation trades at a fixed world price ratio given by the slope of the line CE. If perfect competition prevails within the nation, production occurs at point D. Thus consumption can occur anywhere along CE. All the points on the line CE (except D itself) lie strictly outside the pre-trade set of possible consumption

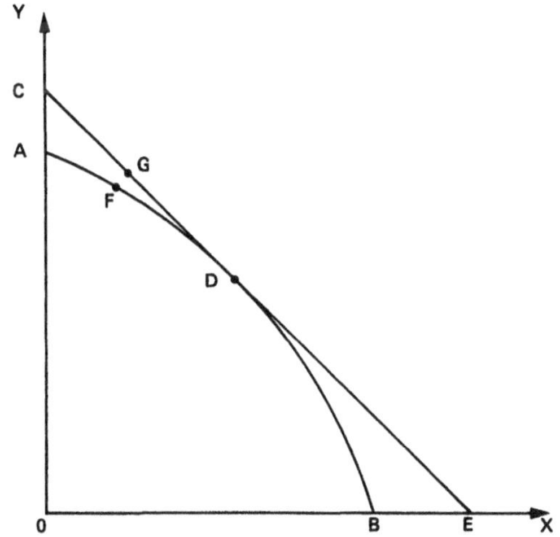


FIGURE 11.1 The gain from trade.

points. In some overall sense, the nation is made better off because its consumption possibilities have increased. But what is meant by the welfare of a nation?

There is no reason why all citizens would prefer any given free-trade allocation (along CE) to any given autarky allocation (along AB). On the contrary, there is a presumption that some citizens prefer autarky: owners of the nation's relatively scarce resource are likely to be made worse off by the opening of international trade. The movement from a point on AB to a point on CE involves changes in the distribution of income which are not made apparent by Figure 11.1; points on CE are not unanimously preferred to points on AB.

One weak statement can be made in comparing pre-trade and post-trade welfare. This is that a post-trade point such as G in Figure 11.1 is preferable to a pre-trade point such as F in that, since greater quantities of both goods are available, it is possible to redistribute income in such a way as to make all citizens better off at G than at F. The gains from trade are large enough so that it is possible for the gainers to compensate the losers and still gain from the opening of trade. Such a result is of limited interest, however, because it says nothing about whether or how such redistributions are to be achieved; the gain from trade is only a potential welfare gain.

The necessity for such inconclusive welfare statements is avoided if it is assumed that individuals' preferences can be aggregated so that there exists a utility function characterizing the level of welfare of the entire nation as a function of total quantities consumed. If this function has the usual properties of utility functions, it can be represented by a set of indifference curves. Then, because free trade generates an expanded set of consumption possibilities, it can be concluded from Figure 11.1 that free trade is preferred to autarky.

It should be stressed that community preference functions are useful not only in normative analysis, but also in positive analysis. For example, if consumption choices are made by individuals according to their own preferences, where on the line CE in Figure 11.1 would the free-trading economy's equilibrium be? Answering such questions is greatly simplified if the outcome of the many individual choices is as if there were a single individual maximizing utility by choosing total consumption quantities; that is, if community indifference curves exist.

The concept of the community preference function is also useful in empirical analysis. Usually, data are available only for aggregates of households. Aggregate consumption data can be represented as resulting from a single optimization problem if a community preference function exists.

Not surprisingly, the conditions under which consumers' choices can be represented by a community preference function are very restrictive: these conditions are discussed in the next two sections.

### 11.3 Existence of a community preference function

If the distribution of income is constrained so that each individual's income is a constant share of total income, and if the individual utility functions satisfy certain restrictions (for example, homothetic but not necessarily identical), then the nation's choices can be represented by a community preference function. This result has little relevance for international economics, however, because the questions to be analyzed (the effects of a change in tariff rates, for example) inevitably involve changes in the distribution of income.

When does a community preference function exist regardless of changes in the distribution of income? A necessary and sufficient condition is that each individual consumes a positive amount of each commodity, and all consumers' Engel curves are parallel straight lines; that is, that the demand function of individual $h$, $h = 1, \ldots, H$, for good i, $i = 1, \ldots, n$, is of the form

$$x_i^h = \alpha_i^h(p) + \beta_i(p)m^h, \tag{11.1}$$

where $m^h$ is the individual's income and $p = (p_1, \ldots, p_n)$ is the vector of prices. With the demand functions (11.1), the marginal propensity to consume a particular good ($\beta_i(p)$) is the same for each individual. Thus taking one dollar of income from one consumer and giving it to another does not change the total quantities purchased; total consumption of each commodity is independent of the distribution of income. Notice that, although each individual's Engel curve has the same slope ($\beta_i(p)$), the Engel curves need not have the same vertical intercept ($\alpha_i^h(p)$, $h = 1, \ldots, H$).

Demand functions of the form (11.1) in turn impose strong restrictions on the utility functions. An indirect utility function shows the maximal utility a consumer can attain as a function of prices and income. Provided all individuals consume a positive amount of each good, it is necessary and sufficient for demand functions to have the form (11.1) that each individual's indirect utility function be of the form

$$u^h = \frac{m^h - a^h(p)}{b(p)} \tag{11.2}$$

for all $h = 1, \ldots, H$, where $a^h$, $h = 1, \ldots, H$, and $b$ are linear homogeneous and concave functions of $p$. The functional form (11.1) is called the *Gorman polar form.* Figure 11.2 depicts a set of indifference curves for such a utility function.

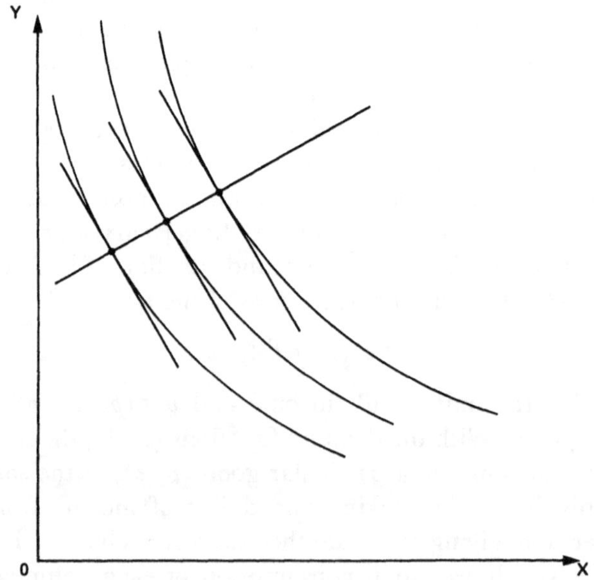


FIGURE 11.2 Indifference curves for the Gorman polar form.

### 11.4* Existence of a community preference function: proof

We show first that the outcome of many individual consumer choices is as if the choices were made by a single consumer if and only if individual demand functions are linear in income. Suppose the demand function of individual $h$ for good $i$ is

$$x_i^h = f_i^h(m^h, p), \tag{11.3}$$

so that total demand $x_i$ is

$$x_i = \sum_{h=l}^{H} f_i^h(m^h, p). \tag{11.4}$$

It is as if this total demand arises from a single consumer optimization problem if total demands depend only on prices and total income; that is, if (11.4) can be reduced to the form

$$x_i = g_i(m, p) \tag{11.5}$$

for all $i = 1, \ldots, n$. For (11.5) to be satisfied, total demand must be independent of the distribution of income; taking $1 from one consumer and giving it to another must not change market demand. Thus the marginal propensities to spend must be constant and the same for all individuals. Thus individual demand functions must be linear in income; (11.3) must be of the form (11.1). Here it must be assumed that each individual's income is large enough so that each individual demand $x_i^h$ found by solving (11.1) is nonnegative. The market demand functions corresponding to (11.1) are

$$x_i = \alpha_i(p) + \beta_i(p)m \tag{11.6}$$

where $\alpha_i(p) \equiv \sum_{h=1}^{H} \alpha_i^h(p)$. Thus we have shown that it is necessary and sufficient for market demands to behave as if they were generated by a single consumer that all individuals' Engel curves be parallel straight lines.

We next prove that individual demand functions are of the linear form (11.1) if and only if utility functions are of the Gorman polar form (11.2). To show first that (11.2) implies (11.1), use Roy's identity:

$$x_i^h = -\frac{\partial u^h / \partial p_i}{\partial u^h / \partial m^h} \tag{11.7}$$

Thus, differentiating (11.2), the demand functions are

$$x_i^h = \left[a_i^h(p) - \frac{b_i(p)}{b(p)}a^h(p)\right] + \frac{b_i(p)}{b(p)}m^h, \tag{11.8}$$

which is of the form (11.1). To show the converse, let $c^h(u^h, p)$ represent individual $h$'s cost function; this shows the minimum expenditure needed to attain the utility level $u^h$ at prices $p$. Write the cost function as

$$c^h(u^h, p) = \gamma^h(p) + \theta^h(u^h, p) \tag{11.9}$$

with $\theta^h(0, p) = 0$. The derivative of the cost function with respect to the price of good $i$ is the compensated demand function for good $i$:

$$x_i^h = c_i^h(u^h, p) = \gamma_i^h(p) + \theta_i^h(u^h, p) \tag{11.10}$$

But substituting (11.9) in (11.1) yields

$$x_i^h = [\alpha_i^h(p) + \beta_i(p)\gamma_i^h(p)] + \beta_i(p)\theta^h(u^h, p) \tag{11.11}$$

Comparing (11.11) and (11.10),

$$\beta_i(p)\theta^h(u^h, p) = \theta_i^h(u^h, p); \tag{11.12}$$

from which there exist functions $\zeta$ and $\eta$ such that

$$\theta^h(u^h, p) = \zeta^h(u^h)\eta^h(p) \tag{11.13}$$

$\zeta$ is a monotonic increasing function, so $u^{h*} = \zeta(u^h)$ is an equivalent utility function. Using this, and substituting (11.13) in (11.9), the cost function is

$$c(u^{h*}, p) = \gamma^h(p) + \eta^h(p)u^{h*} \tag{11.14}$$

Rearranging this cost function, and substituting $m = c$, yields the utility function (11.2). Hence Gorman-polar-form utility is equivalent to linear Engel curves.

### 11.5 Transferable utility in an open economy

Sum together the utility functions (11.2):

$$\sum_{h=1}^{H} u^h = \frac{m - a(p)}{b(p)}, \tag{11.15}$$

where $a(p) \equiv \sum_{h=1}^{H} a^h(p)$. Thus the economy as a whole, with total income $m \equiv \sum_{h=1}^{H} m^h$, behaves as if it were a single consumer with a Gorman-polar-form utility function.

Consider a country which is small, in the sense that the quantities of the commodities it buys and sells on world markets are too small a fraction of total world trade to affect world prices. World prices and the domestic tariff structure determine where on the country's production-possibility frontier production takes place. (Figure 11.1 illustrates this for the no-tariff case.) Production quantities are decided independently of domestic consumers' preferences in the case of the small country. Given production quantities, the country's total income, evaluated at world prices, is determined; denote this by $m$. Then equation (11.15) shows that the country's utility-possibility frontier is linear. Gorman-polar-form utility implies that utility is transferable within the small, open economy.

Conversely, suppose utility is transferable within a small, open economy. The linearity of the utility-possibility frontier means that a change in the distribution of income must leave total utility unchanged. This in turn requires that each individual consumer have the same constant marginal utility of income. Given that the smallness of the economy makes prices exogenous, this implies (by the analysis of Sections 11.3 and 11.4) that utility is of the Gorman polar form.

Hence utility is transferable within a small, open economy if and only if there exists a community preference function.

## 11.6 Countries as coalitions

The reason for the usefulness of community preference functions in international economics is essentially the same as the reason for the usefulness of transferable utility in game theory. The analysis is considerably simplified if the aggregate allocation of resources can be determined independently of the distribution of the ownership of resources. Gorman-polar-form utility ensures this independence.

Although the utility function (11.2) is special, it is equivalent to the linearity of Engel curves. Linear Engel curves may be satisfactory local approximations to arbitrary Engel curves. This does not justify assuming the existence of community preference functions, however (even for analyzing small changes) because, in addition to being linear, the Engel curves must be parallel for all individuals. Thus the conditions for existence of community preference functions (or transferable utility) are stringent.

In much of international economics, the country is treated as the unit of analysis; it is as though each country consisted of a single decision-maker. This is reminiscent of the characteristic-function approach to the theory of cooperative games, in which the unit of analysis is the coalition. It has been shown that each of these approaches is valid if individual consumers have Gorman-polar-form utility functions. This suggests a game-theoretic interpretation of international economics. Individual consumers are restricted in their opportunities to form coalitions; instead, the coalition structure is imposed exogenously, with each individual assigned to one coalition, that is, one country. Within each coalition/country, there are laws of contract, so that games may be played cooperatively. However, relations between different coalitions/countries must proceed as noncooperative games.

### 11.7 Bibliographic notes

On the role of community preferences in international economics, see Chipman [26]. On the role of transferable utility in game theory, see Luce and Raiffa [96] and Shubik [146].

The relationship between Gorman-polar-form utility functions and commmunity-preference functions was derived by Gorman [50]; see also Gorman [52]. If restrictions are placed on the distribution of income, or if the government redistributes income so as to maximize some social-welfare function, then more general utility functions are consistent with the existence of community preference functions: see Woodland [164], Chapter 6.

Bergstrom and Varian [11] showed that an exchange economy can be modelled as a transferable-utility game if utility functions are of the Gorman polar form.

## 12. FURTHER TOPICS

The list of questions in international economics to which game theory is applicable has not been exhausted. In this section, a few more examples of game-theoretical analysis are briefly noted.

### 12.1 The political economy of tariffs

In most of the preceding models, the country was portrayed as a monolithic agent. This ignores possible conflicts of interest among citizens within a country. For example, in an economy with two factors of production, imposing a tariff raises the real income of owners of the relatively scarce resource and lowers the real income of owners of the relatively abundant resource. Casual empiricism suggests that a country's international economic policy is often in fact designed to redistribute income domestically. A positive theory of international trade, therefore, might need to take account of political economy.

Why would a small open economy impose tariffs, given that those tariffs generate an aggregate welfare loss? One answer presumes that the political process of making a decision to introduce or to remove a tariff is costly. If the potential gainers from the tariff are concentrated and each gainer individually gains much from the tariff, while the potential losers are diffuse and their *per capita* losses are small, then the gainers will be able to muster more resources to support their cause. The political decision will be made in favor of the tariff, even though, in a world of costless decision-making, there could be designed a scheme of transfer payments such that everyone would be better off under free trade.

To formalize this argument, consider (following Brock and Magee[21,22]) a model of voluntary contributions to a political lobby seeking the enactment of a tariff. Denote by $s_i$ the increase in the present value of individual $i$'s income should the tariff be enacted. Let $g_i$ be the monetary contribution of agent $i$ to the lobby. The larger is the sum of contributions, $\sum_{j=1}^{n} g_j$, the higher is the probability, $\pi(\sum_{j=1}^{n} g_j)$, that the lobby will succeed in having the tariff legislation passed (though this also depends upon the activities of any counter-lobbies, not discussed here). The expected net benefit to individual $i$ from contributing to the lobby is

$$b_i = \pi\left(\sum_{j=1}^{n} g_j\right)s_i - g_i \tag{12.1}$$

Note the free-rider problem inherent in the lobbying activity: net benefit increases in total contributions, not own contribution alone, but decreases in own contribution.

Maximizing the net benefit, allowing for the possibility of a zero contribution,

$$\pi'\left(\sum_{j=1}^{n} g_j\right)s_i - 1 \leqslant 0, \text{ with} = \text{only if } g_i > 0. \tag{12.2}$$

The Nash equilibrium of this game is found by solving simultaneously n conditions of the form (12.2).

Clearly if an individual's stake in the tariff, $s_i$, is positive but not very high, then (12.2) indicates that he will contribute nothing. It is possible that no one contributes, so that the lobby is not formed. This depends only on the size of individual stakes $s_i$, and not on the total gains $\sum_{j=1}^{n} s_j$. If all the individuals have a small stake in the outcome, the group will not be able to overcome the free-rider problem and act cohesively, regardless of how large the total potential gains are.

Consider now the possibility of an interior solution to (12.2). For given total contributions, at most one value of $s_i$ satisfies (12.2) as an equation. If all individuals have different stakes in the outcome, only the individual with the largest stake will contribute to the lobbying. Everyone but the individual or individuals with the most to gain from the tariff will be a free rider. Nevertheless, provided some individual's stake $s_i$ is large enough that $\pi'(0)s_i > 1$, the lobbying will take place.

The conclusion that only the largest potential gainer contributes would be changed in an empirically more reasonable direction if the net benefit function (12.1) were amended to include the possibility that each individual benefits from his contribution not only indirectly, in increasing the probability of the lobby's success, but also directly, perhaps because the contribution enables the contributor to move the lobby's activities closer to his particular interests. The individual's net benefit function would be of the form

$$b_i = \pi\left(\sum_{j=1}^{n} g_j\right) s_i + \beta_i(g_i) - g_i, \qquad (12.4)$$

with $\beta_i'(g_i) > 0$. The larger is the direct benefit from participating in the lobby, $\beta_i(g_i)$, the more likely it is that the individual will participate.

## 12.2 Strategic macroeconomic policies

In the 1930's, countries tried to export their unemployment problem by manipulating their exchange rates. Politicians have been more circumspect recently, but it is clear that there still exist interactions among different nations' macroeconomic policies: during 1983, for example, European countries blamed their unemployment on the high United States interest rates and budget deficit.

Consider (following Hamada [55, 56]) a simple international monetary model. There are n countries operating under fixed exchange rates. There is a common world price level and so a common world inflation rate $\rho$. Each country seeks to control the inflation rate and its balance of payments $z_i$ according to a utility function $u^i(\rho, z_i)$; there is a target rate of inflation $\rho^*$ and balance of payments $z_i^*$ such that

$$u_1^i \gtreqless 0 \text{ as } \rho \gtreqless \rho^*,$$

$$u_2^i \gtreqless 0 \text{ as } z_i \gtreqless z_i^*. \qquad (12.5)$$

Each government's policy instrument is the rate of excess credit expansion $x_i$. The variables are related by:

$$z_i = \rho - x_i, \ i = 1, \ldots, n \qquad (12.6)$$

Let $m_i$ be the equilibrium quantity of money in country $i$ and $m = \sum_{j=1}^{n} m_j$. Let $w_i \equiv m_i/m$. Summing the $n$ equations (12.6),

$$\rho = \sum_{j=1}^{n} w_j x_j + \sum_{j=1}^{n} w_j z_j. \qquad (12.7)$$

Thus country i's welfare can be written as

$$u^i = u^i\left(\sum_{j=1}^{n} w_j x_j + \sum_{j=1}^{n} w_j z_j, \sum_{j=1}^{n} w_j x_j + \sum_{j=1}^{n} w_j z_j - x_i\right), \qquad (12.8)$$

exhibiting the interdependencies among the countries' monetary policies. Differentiating with respect to $x_i$, and rearranging,

$$w_i u_1^i = (1 - w_i) u_2^i, \; i = 1, \ldots, n. \qquad (12.9)$$

Equations (12.9) define the Nash equilibrium. Summing the $n$ equations:

$$\sum_{j=1}^{n} w_j u_1^j = \sum_{j=1}^{n} (1 - w_j) u_2^j. \qquad (12.10)$$

The set of Pareto optimal points is found by maximizing $\sum_{j=1}^{n} \omega_j u^j$ for arbitrary weights $\omega_i$, $0 < \omega_i < 1$, $\sum_{j=1}^{n} \omega_j = 1$. It can be shown that, at Pareto optimum,

$$\sum_{j=1}^{n} \omega_j u_1^j = 0 \qquad (12.11)$$

Comparing (12.10) and (12.11), the Nash equilibrium cannot be Pareto optimal. The noncooperative equilibrium may involve either too much or too little inflation.

The Pareto inferiority of the noncooperative equilibrium indicates a role for international negotiations to attempt to achieve some agreement among nations on preferred macroeconomic policies.

### 12.3 State trading

The term "countertrade" has been coined to describe a transaction in which two or more otherwise independent transactions are formally linked, usually when the trading agents are governments. Countertrade consists in part of barter. For example, a government buying goods from a foreign supplier might require the supplier to reciprocate by purchasing some goods made in that country. Countertrade arrangements are frequently used by developing countries or centrally planned countries. Countertrade is quantitatively an important phenomenon: it has been estimated that between one-fifth and one-third of world trade takes place in this way.

The model of small buyers and sellers trading at market-determined prices is not relevant to this kind of trade. Each transaction, rather than being one of many essentially identical transactions, is unique. Each transaction takes place between a particular pair of countries. The "price" is determined by bilateral negotiations. Game theory is obviously relevant to the modelling of such a phenomenon.

A related issue is that governments often favor local suppliers in the sealed-bid tendering for government contracts. One way in which this is done is by offering a price preference based on local content; for example, in United States defense contracts, a local bidder will win the contract if his bid is no more than 50 percent higher than the lowest bid of a foreign firm. The welfare consequences of such a policy are more subtle than the welfare consequences of, say, a tariff. The government is the sole buyer; it therefore has some power to dictate the terms of the purchase. There are typically only a few sellers, so that the sellers behave strategically towards each other. There is incomplete information. Because bids are sealed, each firm must decide how low to bid without knowing the others' bids. The firms have different expected costs; these expected costs cannot be observed by the government. Imposing a procurement preference changes the extent of competition in bidding: it tends to raise the bids of the domestic firms and to lower the bids of the foreign firms. It may raise or lower the expected cost of the project to the government. There is no basis for any presumption that the welfare-maximizing policy would be to eliminate such procurement preferences.

## 12.4 Trade disruption

International trade sometimes is carried out under the threat that at any time one country might cut off trade. For example, a foreign-owned resource-extracting firm may face the threat of being nationalized by the host country. Or an importing country might react to what it sees as unduly an aggressive policy on the part of the exporting country by invoking quantitative restrictions on imports. Or a monopolistic exporter might seek to increase his market power by threatening a trade embargo. The classical national-defense argument for tariffs is about trade disruption.

In many cases, the probability of disruption is endogenous. Thus the timing or extent of the quotas imposed by an importing country might vary with the exporting country's choice of actions. This is a problem in dynamic game theory. The exporting country chooses the best time-path of its exports given what it knows about how the importing country will respond. The importing country, on the other hand, given its objectives, chooses a decision rule about when and how extensively to invoke quotas as a function of the time-path of its imports.

### 12.5 Bibliographic notes

The analysis of the political economy of tariffs in Section 12.1 was a simplified version of part of that of Brock and Magee [21, 22]; see also Bhagwati [12] and Stigler [152]. An early study of the political economy of tariffs is Kindleberger [83]. For an analysis of collusion among factor owners using the concept of the core, see Hansen and Jaskold-Gabszewicz [58].

The simple macroeconomic model of Section 12.2 is due to Hamada [55, 56]. Nash equilibria for other games of international macroeconomic interactions were analyzed by Johansen [67] and Manning [97].

For a game-theoretic analysis of government procurement preferences, see McAfee and McMillan [106]. For institutional details on state trading and international barter arrangements see Adkins [2], Kostecki [86], and Viner [158].

For models of trade disruption, see Bhagwati and Srinivasan [14], Crawford, Sobel and Takahashi [30], and Kemp and Ohta [80].

(*University of Western Ontario*)

## REFERENCES

1. Abreu, D. Repeated games with discounting: a general theory and an application to oligopoly, Ph.D. thesis, Princeton University, 1983.
2. Adkins, D. L. Countertrade, clearing arrangements, reciprocity and other instruments of the new bilateralism in international trade, mimeo, Office of Technology Assessment, U.S. Congress, 1983.

3. Allen, D. L., Tariff games, *Applied Game Theory*, ed. by S. J. Brams, A. Schotter and G. Schwödiauer. Wuerzburg: Physica-Verlag, 1979.
4. Anderson, Lee G. Optimum economic yield on an internationally utilized common property resource, *Fishery Bulletin*, **73** (1975), 51–56.
5. Arrow, K. J. The property rights doctrine and demand revelation under incomplete information, *Economics and Human Welfare: Essays in Honor of Tibor Scitovsky*, ed. by M. J. Boskin. New York: Academic Press, 1979.
6. Arrow, K. J., and F. H. Hahn. *General Competitive Analysis*. San Francisco: Holden-Day, 1971.
7. Aumann, Robert J. Survey of repeated games, *Essays in Game Theory and Mathematical Economics in Honor of Oskar Morgenstern*, ed. by V. Böhm and H. H. Nachtkamp. Mannheim: Bibliographisches Institut, 1981.
8. Benoit, J. P., and Krishna, V. Finitely repeated games, Presented at conference on Strategic Behavior and the Theory of the Firm, University of Western Ontario, 1984.
9. Bergstrom, Theodore C. On capturing oil rents with a national excise tax, *American Economic Review*, **72** (1982), 195–201.
10. Bergstrom, Theodore, Lawrence Blume, and Hal Varian. On the private provision of public goods, Discussion Paper No. C-56, University of Michigan, 1983.
11. Bergstrom, Theodore C., and Hal R. Varian. When do market games have transferable utility? Discussion Paper No. C-53, University of Michigan, 1983.
12. Bhagwati, J. N. *Import Competition and Response*. Chicago: University of Chicago Press, 1982.
13. Bhagwati, J. N., R. A. Brecher, and T. Hatta. The generalized theory of transfers and welfare. Bilateral transfers in a multilateral world, *American Economic Review*, **83** (1983), 606–618.
14. Bhagwati, J. N., and T. N. Srinivasan. Optimal trade policy and compensation under endogenous uncertainty: The phenomenon of market disruption, *Journal of International Economics*, **6** (1976), 317–336.
15. Bolle, F.: The efficient use of a non-renewable common-pool resource is possible but unlikely, *Zeitschrift für Nationalökonomie*, **40** (1980), 391–397.
16. Brander, James A.: Intra-industry trade in identical commodities, *Journal of International Economics*, **11** (1981), 1–14.
17. Brander, James A., and Slobodan Djajic: Rent-extracting tariffs and the management of exhaustible resources, *Canadian Journal of Economics*, **16** (1983), 288–298.
18. Brander, James A., and Barbara J. Spencer. Tariffs and the extraction of foreign monopoly rent, *Canadian Journal of Economics*, **14** (1981), 371–389.
19. Brander, James A., and Barbara J. Spencer, Tariff protection and imperfect competition, *Monopolistic Competition and International Trade*, ed. by H. Kierzkowski. Oxford: Oxford University Press, 1984.
20. Brandt, Willy, *et al. North-South: A programme for survival*. London: Pan, 1980.
21. Brock, William A., and Stephen P. Magee. The economics of special interest politics: The case of the tariff, *American Economic Review: Papers and Proceedings*, **68** (1978), 246–250.
22. Brock, William A., and Stephen P. Magee. Tariff setting in a democracy, *Current Issues in International Commercial Policy and Economic Diplomacy*, ed. by J. Black and B. Hindley. London: Macmillan, 1979.

23. Chiarella, Carl. Trade between resource-poor and resource-rich economies as a differential game, *Exhaustible Resources, Optimality, and Trade*, ed. by M. C. Kemp and N. V. Long. Amsterdam: North-Holland, 1980.
24. Chiarella, Carl, Murray C. Kemp, Ngo van Long, and Koji Okuguchi. On the economics of international fisheries, *Essays in the Economics of Exhaustible Resources*, ed. by M. C. Kemp and N. V. Long. Amsterdam: North-Holland, 1984.
25. Chichilnisky, Graciela. Basic goods, the effects of commodity transfers and the international economic order, *Journal of Development Economics*, **7** (1980), 505–520.
26. Chipman, John S. A survey of the theory of international trade, *Econometrica*, **33** (1965), 477–519, 685–760, and **34** (1966), 18–76.
27. Clark, Colin W. Restricted access to common-property fishing resources: A game-theoretic analysis, *Dynamic Optimization and Mathematical Economics*, ed. by Pan-Tai Liu. New York: Plenum Press, 1980.
28. Cournot, Augustin. *Recherches sur les Principes Mathématiques de la Théorie des Richesses*. Paris: L. Machette, 1838. Translated as *Mathematical Principles of the Theory of Wealth*. New York: Kelley, 1971.
29. Crawford, Vincent P. Dynamic contract theory, with applications to country foreign borrowing strategies, Discussion Paper No. 84–9, University of California, San Diego, 1984.
30. Crawford, Vincent P., Joel Sobel, and Ichiro Takahashi. Bargaining, strategic reserves, and international trade in exhaustible resources, Discussion Paper 80–30, University of California, San Diego, 1982.
31. Dam, Kenneth W. *The GATT: Law and International Economic Organization*. Chicago: University of Chicago Press, 1970.
32. Dasgupta, Partha, Richard Gilbert, and Joseph Stiglitz. Strategic considerations in invention and innovation: The case of natural resources, *Econometrica*, **51** (1983), 1439–1448.
33. Debreu, Gerard. *Theory of Value*. New Haven: Yale University Press, 1959.
34. Dixit, Avinash K. International trade policy for oligopolistic industries, *Economic Journal Conference Papers*, **94** (1984), 1–16.
35. Dixit, Avinash K., and Gene M. Grossman, Targeted export promotion with several oligopolistic industries, Discussion Paper No. 71, Woodrow Wilson School, Princeton University, 1984.
36. Dixit, Avinash K., and Albert Kyle. On the use of trade restrictions for entry promotion and deterrence, Discussion Paper No. 56, Woodrow Wilson School, Princeton University, 1983.
37. Dixit, Avinash K., and Victor Norman. *The Theory of International Trade*. Cambridge: Cambridge University Press, 1980.
38. Dubey, Pradeep. Nash equilibria of market games: Finiteness and efficiency, *Journal of Economic Theory*, **22** (1980), 363–376.
39. Eaton, Jonathon, and Mark Gersovitz. Debt with potential repudiation: theoretical and empirical analysis, *Review of Economic Studies*, **48** (1981), 289–310.
40. Eaton, Jonathon, and Gene M. Grossman, Optimal trade and industrial policy under oligopoly, Discussion Paper No. 59, Woodrow Wilson School, Princeton University, 1983.
41. Eckbo, Paul Lee. *The Future of World Oil*. Cambridge, Mass.: Ballinger, 1976.
42. Edgeworth, F. Y. *Mathematical Psychics*. London: Kegan, Paul, 1881.

43. Eswaran, Mukesh, and Tracy Lewis. Appropriability and the extraction of a common property resource, *Economica* **51** (1984): 393–400.
44. Friedman, James W. *Oligopoly and the Theory of Games*. Amsterdam: North-Holland, 1977.
45. Friedman, Milton and Rose Friedman. *Free to Choose*. London: Penguin, 1980.
46. Fudenberg, Drew, and David Levine. Subgame-perfect equilibria of finite- and infinite-horizon games, *Journal of Economic Theory*, **31** (1983), 251–268.
47. Gallini, Nancy, Tracy Lewis, and Roger Ware. Strategic timing and pricing of a substitute in a cartelized resource market, *Canadian Journal of Economics*, **16** (1983), 429–446.
48. Gately, Dermot. Sharing the gains from customs unions among less developed countries, *Journal of Development Economics*, **1** (1974), 213–233.
49. Gately, Dermot. OPEC pricing and output decisions: A partition function approach to OPEC stability, *Applied Game Theory*, ed. by S. Brams, A. Schotter, and G. Schwodiauer. Wuerzburg: Physica-Verlag, 1979.
50. Gorman, W. M. Community preference fields, *Econometrica*, **21** (1953), 63–80.
51. Gorman, W. M. Tariffs, retaliation, and the elasticity of demand for imports, *Review of Economic Studies*, **33** (1958), 133–162.
52. Gorman, W. M. Tricks with utility functions, *Essays in Economic Analysis*, ed. by M. J. Artis and A. R. Nobay. Cambridge: Cambridge University Press, 1976.
53. Green, E. J. Non-cooperative price taking in large dynamic markets, *Journal of Economic Theory*, **22** (1980), 155–182.
54. Grinols, Earl L. An extension of the Kemp-Wan theorem on the formation of customs unions, *Journal of International Economics*, **11** (1981), 259–266.
55. Hamada, K. A strategic analysis of monetary interdependence, *Journal of Political Economy*, **84** (1976), 677–700.
56. Hamada, K. A game-theoretic approach to international monetary confrontations, *Applied Game Theory*, ed. by S. J. Brams, A. Schotter, and G. Schwödiauer, Wuerzburg: Physica-Verlag, 1979.
57. Hamilton, Bob, and John Whalley. Optimal tariff calculations in alternative trade models and some possible implications for current world trading arrangements, *Journal of International Economics*, **15** (1983), 323–348.
58. Hansen, Terje, and Jean Jaskold-Gabszewicz. Collusion of factor owners and distribution of factor output, *Journal of Economic Theory*, **4** (1972), 1–18.
59. Harris, Richard, with David Cox. *Trade, Industrial Policy, and Canadian Manufacturing*. Toronto: Ontario Economic Council, 1984.
60. Harsanyi, John C. Game theory and the analysis of international conflict, *Australian Journal of Politics and History*, **11** (1965), 292–304.
61. Hellwig, Martin F. A model of borrowing and lending with bankruptcy, *Econometrica*, **45** (1972), 1879–1906.
62. Hnyilicza, E., and R. S. Pindyck. Pricing policies for a two-part exhaustible resource cartel: The case of OPEC, *European Economic Review*, **8** (1976), 136–154.
63. Horstmann, Ignatius, and James R. Markusen. Strategic investments, firm specific capital, and the development of multinationals, mimeo, University of Western Ontario. 1983.

64. Horwell, D. J. Optimum tariffs and tariff policy, *Review of Economic Studies*, **33** (1966), 147–158.
65. Hwang, Hong. Intra-industry trade and oligopoly: A conjectural variations approach, *Canadian Journal of Economics*, **17** (1984), 126–137.
66. Ichiishi, Tatsuro, and Raymond Riezman, Customs union formation: A cooperative game-theoretical analysis, mimeo, University of Iowa, 1983.
67. Johansen, Leif. A note on the possibility of an international equilibrium with low levels of activity, *Journal of International Economics*, **13** (1982), 257–265.
68. Johansen, Leif. On the status of the Nash type of non-cooperative equilibrium in economic theory, *Scandinavian Journal of Economics*, **84** (1982), 421–441.
69. Johnson, H. G. Optimum tariffs and retaliation, *Review of Economic Studies*, **21** (1953), 142–153. Reprinted with amendments in *International Trade and Economic Growth*. Cambridge: Harvard University Press, 1961.
70. Jones, Ronald W. The transfer problem in a three-agent setting, *Canadian Journal of Economics*, **17** (1984), 1–14.
71. Kaldor, N. A note on tariffs and the terms of trade, *Economica*, N.S., **7** (1940), 377–380.
72. Kawasaki, Seiichi, John McMillan, and Klaus F. Zimmermann. Inventories and price inflexibility, *Econometrica*, **51** (1983), 599–610.
73. Kemp, Murray, C. *The Pure Theory of International Trade*. Englewood Cliffs, N.J.: Prentice-Hall, 1964.
74. Kemp, Murray, C. *A Contribution to the General Equilibrium Theory of Preferential Trading*. Amsterdam: North-Holland, 1969.
75. Kemp, Murray C. *The Pure Theory of International Trade and Investment*. Englewood Cliffs, N.J.: Prentice-Hall, 1969.
76. Kemp, Murray C. A note on the theory of international transfers, *Economics Letters*, **14** (1984), 259–262.
77. Kemp, Murray, C., and Shoichi Kojima. The welfare economics of foreign aid, *Trends in Contemporary Microeconomics*, ed. by G. R. Feiwel. London: Macmillan, 1982.
78. Kemp, Murray C., and Ngo van Long. The interaction of resource-poor and resource-rich economies, *Australian Economic Papers*, **18** (1979), 258–267. Reprinted in *Exhaustible Resources, Optimality, and Trade*, ed. by M. C. Kemp and N. V. Long. Amsterdam: North-Holland, 1980.
79. Kemp, Murray C., and Ngo van Long. Resource extraction under conditions of common access, *Exhaustible Resources, Optimality, and Trade*, ed. by M. C. Kemp and N. V. Long. Amsterdam: North-Holland, 1980.
80. Kemp, Murray C., and Hiroshi Ohta. The optimal level of exports under threat of foreign import restriction, *Canadian Journal of Economics*, **11** (1978), 720–725.
81. Kemp, Murray C., and Henry Y. Wan. An elementary proposition concerning the formation of customs unions, *Journal of International Economics*, **6** (1976), 95–97.
82. Keynes, J. M. The international control of raw materials, *Journal of International Economics*, **4** (1974), 299–315.
83. Kindleberger, C. P. Group behavior and international trade, *Journal of Political Economy*, **59** (1951), 30–46.
84. Kirman, A. P., and M. J. Sobel, Dynamic oligopoly with inventories, *Econometrica*, **42** (1974), 279–287.

85. Kletzer, Kenneth M. Asymmetries of information and LDC borrowing with sovereign risk, *Economic Journal*, **94** (1984), 287–307.
86. Kostecki, M. M. ed. *State Trading in International Markets*. New York: St. Martins, 1982.
87. Kreps, D. M., P. Milgrom, J. Roberts, and R. Wilson. Rational cooperation in the finitely repeated prisoners' dilema, *Journal of Economic Theory*, **27** (1982), 245–252.
88. Krishna, Kala. Trade restrictions as facilitating practices, Discussion Paper No. 55; Woodrow Wilson School, Princeton University, 1983.
89. Krugman, Paul. Import protection as export promotion: International competition in the presence of oligopoly and economies of scale, *Monopolistic Competition and International Trade*, ed. by H. Kierzkowski. Oxford: Oxford University Press, 1984.
90. Kuga, Kiyoshi. Tariff retaliation and policy equilibrium, *Journal of International Economics*, **3** (1973), 351–366.
91. Lawrence, Colin, and Victor Levy. On sharing the gains from international trade: The political economy of oil consuming nations and oil producing nations, *International Economic Review*, **23** (1982), 711–722.
92. Lee, Tom K. Bilateral trade, dynamic bargaining and nonrenewable resources, *Journal of International Economics*, **14** (1983), 1–10.
93. Léonard, D., and R. Manning. Advantageous reallocations: A constructive example, *Journal of International Economics*, **15** (1983), 291–296.
94. Levhari, D., and L. J. Mirman, The great fish war: An example using a dynamic Cournot-Nash solution, *Bell Journal of Economics*, **11** (1980), 322–334.
95. Lovasy, Gertrud. *International Cartels*. New York: United Nations, 1946.
96. Luce, R. Duncan, and Howard Raiffa. *Games and Decisions*. New York: Wiley, 1957.
97. Manning, R. A Nash-Cournot model of taxation of international capital movements, *Economic Record*, **50** (1974), 113–118.
98. Mantel, Rolf R., and Ana M. Martirena-Mantel. Integración ecónomica, distribution del ingreso y consumo: Una nueva racionalidad para la integración, *El Trimestre Económico*, **42** (1975), 631–669. Translated as "Economic integration, income distribution, and consumption: A new rationale for economic integration, Paper presented at International Conference on Income, Consumption and Prices, Hamburg, 1973.
99. Maskin, Eric, and David Newbery. Rational expectations with market power: The paradox of the disadvantageous tariff on oil, mimeo, Massachusetts Institute of Technology, 1978.
100. Markusen, James R. Cooperative control of international pollution and international common property resources, *Quarterly Journal of Economics*, **89** (1975), 618–632.
101. Markusen, James R. Production and trade from international common property resources, *Canadian Journal of Economics*, **9** (1976), 309–319.
102. Markusen, James R. The distribution of gains from bilateral tariff reductions, *Journal of International Economics*, **11**, (1981), 553–572.
103. Markusen, James R. Trade and the gains from trade with imperfect competition, *Journal of International Economics*, **11** (1981), 531–551.
104. Mayer, W. Theoretical considerations on negotiated tariff adjustments, *Oxford Economic Papers*, **33** (1981), 135–153.

105. McLennan, Andrew. A noncooperative definition of two person bargaining, Working Paper No. 8303, University of Toronto.
106. McAfee, R. Preston, and John McMillan. Discrimination in auctions, Discussion Paper No. 85–7, University of California, San Diego, 1985.
107. McMillan, John. Collusion, competition, and conjectures, *Canadian Journal of Economics*, **17** (1984), 788–805.
108. McMillan, John, and Ewen McCann. Welfare effects in customs unions, *Economic Journal*, **91** (1981), 697–703.
109. McMillan, John and Hans-Werner Sinn. Oligopolistic extraction of a common-property resource: Dynamic Equilibria, *Essays in the Economics of Exhaustible Resources*, ed. by M. C. Kemp and N. V. Long. Amsterdam: North-Holland, 1984.
110. Melvin, James R., and Robert Warne. Monopoly and the theory of international trade, *Journal of International Economics*, **3** (1973), 117–134.
111. Munro, Gordon, R. The optimal management of transboundary renewable resources, *Canadian Journal of Economics*, **12** (August 1979), 355–376.
112. Nash, J. F. The bargaining problem, *Econometrica*, **18** (1950), 155–162.
113. Nash, J. F. Non-cooperative games, *Annals of Mathematics*, **54** (1951), 286–295.
114. Negishi, Takashi. Monopolistic competition and general equilibrium, *Review of Economic Studies*, **28** (1961), 196–201.
115. Newbery, D. M. G. Commodity price stabilization in imperfectly competitive markets, Discussion Paper No. 32, Development Research Center, World Bank, 1981.
116. Newbery, D. M. G., and J. E. Stiglitz. Optimal commodity stockpiling rules, *Oxford Economic Papers*, **34** (1982), 403–427.
117. Nichols, A. L., and R. J. Zeckhauser. Stockpiling strategies and cartel prices, *Bell Journal of Economics*, **8** (1977), 66–96.
118. Novshek, William. On the existence of Cournot equilibrium, *Review of Economic Studies* **52** (1985): 85–98.
119. Nurkse, Ragnar. Trade fluctuations and buffer policies of low-income countries, *Kyklos*, **11** (1958), 141–154.
120. Ohyama, M. Trade and welfare in general equilibrium, *Kieo Economic Studies*, **9** (1972), 37–73.
121. Olson, Mancur, and Richard Zeckhauser. An economic theory of alliances, *Review of Economics and Statistics*, **48** (1966), 266–279.
122. Ordover, J. A., and R. D. Willig. Perspectives on mergers and world competition, Paper Presented at NBER Conference on Strategic Behavior in International Trade, 1983.
123. Otani, Yoshihiko. Strategic equilibrium of tariffs and general equilibrium, *Econometrica*, **48** (1980), 643–662.
124. Otani, Yoshihiko. Efficient strategy and strategic equilibrium of tariffs, Discussion Paper No. 97, University of Kansas, 1981.
125. Owen, Guillermo. *Game Theory*. Philadelphia: Saunders, 1968.
126. Panchamukhi, V. R. A theory for optimal tariff policy, *Indian Economic Journal*, **9** (1961), 178–198.
127. Pindyck Robert S. The cartelization of world commodity markets, *American Economic Review: Papers and Proceedings*, **69** (1979), 154–158.

128. Porter, Robert H. Optimal cartel trigger price strategies, *Journal of Economic Theory*, **29** (1983), 313–338.
129. Porter, Robert H. Tariff policies in a small open spatial economy, *Canadian Journal of Economics*, **17** (1984), 270–282.
130. Quyen, N. V. The exploitation of a nonrenewable resource under imperfect competition, unpublished Ph.D. Thesis, University of Western Ontario, 1984.
131. Radner, R. Collusive behavior in non-cooperative epsilon-equilibria of oligopolies with long but finite lives, *Journal of Economic Theory*, **22** (1980), 136–154.
132. Raiffa, H. *The Art and Science of Negotiation*. Cambridge, Mass.: Harvard University Press, 1982.
133. Reinganum, Jennifer, F., and Nancy L. Stokey. Oligopoly extraction of a nonrenewable, common property resource: The importance of the period of commitment in dynamic games, Discussion Paper No. 508, Northwestern University, 1981.
134. Riezman, Raymond. Tariff retaliation from a strategic viewpoint, *Southern Economic Journal*, **48** (1982), 583–593.
135. Roberts, J., and H. Sonnenschein. On the foundations of monopolistic competition, *Econometrica*, **45** (1977), 101–113.
136. Robinson, Joan. Beggar-my-neighbour remedies for unemployment, *Readings in the Theory of International Trade*. Homewood, Ill.: Irwin, 1950.
137. Robson, Arthur J. OPEC versus the west: A robust equilibrium, *Journal of Environmental Economics and Management*, **10** (1983), 18–34.
138. Robson, Arthur J., and John McMillan. Dynamic duopoly under demand uncertainty, *Canadian Journal of Economics*, **17** (1984): 695–698.
139. Rodriguez, Carlos Alfredo. The non-equivalence of tariffs and quotas under retaliation, *Journals of International Economics*, **4** (1974), 295–298.
140. Rubenstein, Ariel. Equilibrium in supergames with the overtaking criterion, *Journal of Economic Theory*, **21** (1979), 1–9.
141. Schelling, Thomas C. *The Strategy of Conflict*. Cambridge: Harvard University Press, 1960.
142. Scitovsky, T. A reconsideration of the theory of tariffs, *Review of Economic Studies*, **9** (1941), 89–110. Reprinted in *Readings in the Theory of International Trade*. Homewood, Ill.: Irwin, 1950.
143. Segal, D. On making customs unions fair: An East African example, *Yale Economic Essays*, **10** (1970), 115–160.
144. Selten, R. Reexamination of the perfectness concept for equilibrium points in extensive games, *International Journal of Game Theory*, **4** (1975), 25–55.
145. Shaked, Avnar, and John Sutton. Natural oligopolies and international trade, *Monopolistic Competition and International Trade*, ed. by H. Kierzkowski. Oxford: Oxford University Press, 1984.
146. Shubik, Martin. *Game Theory in the Social Sciences*. Cambridge: MIT Press, 1982.
147. Sinn, Hans-Werner. Common property resources, storage facilities and ownership structures: A Cournot model of the oil market, *Economica*, to appear.
148. Sobel, Matthew, J. Myopic solutions of Markov decision processes and stochastic games, *Operations Research*, **29** (1981), 995–1009.

149. Spencer, Barbara, and James A. Brander. International R & D rivalry and industrial strategy, *Review of Economic Studies*, **50** (1983), 707–722.
150. Starr, A. W., and Y. C. Ho. Nonzero-Sum differential games, *Journal of Optimization Theory and Applications*, **3** (1969), 184–206.
151. Starr, A. W., and Y. C. Ho. Further properties of nonzero-sum differential games, *Journal of Optimization Theory and Applications*, **3** (1969), 207–219.
152. Stigler, George J. Free riders and collective action: An appendix to theories of economic regulation, *Bell Journal of Economics*, **5** (1974), 359–365.
153. Stocking, George W., and Myron W. Watkins. *Cartels in Action.* New York: Twentieth Century Fund, 1946.
154. Sweeney, Richard James, Robert D. Tollison, and Thomas D. Willett. Market failure, the common-pool problem, and ocean resource exploitation, *Journal of Law and Economics*, **17** (1974), 179–192.
155. Thursby, Marie, and Richard Jensen. A conjectural variation approach to strategic tariff equilibria, *Journal of International Economics*, **14** (1983), 145–162.
156. Tower, Edward. The optimal quota and retaliation, *Review of Economic Studies*, **42** (1975), 623–630.
157. Ulph, A. M., and D. T. Ulph. International monopoly-monopsony power over oil and capital, Discussion Paper No. 19, University College London, 1982.
158. Viner, J. *Trade Relations Between Free-Market and Controlled Economies*. Geneva: League of Nations, 1943.
159. Von Neumann, J., and O. Morgenstern. *The Theory of Games and Economic Behavior*. Princeton: Princeton University Press, 2nd ed., 1947.
160. Wan, Henry Y. A contribution to the theory of trade warfare, Unpublished Doctoral Thesis, Massachusetts Institute of Technology, 1961.
161. Warr, Peter G. The private provision of a public good is independent of the distribution of income, *Economics Letters*, **13** (1983), 207–211.
162. Weymark, J. A. Welfare optimal tariff revenues and maximum tariff revenues, *Canadian Journal of Economics*, **13** (1980), 615–631.
163. Wijkman, Per Magnus. UNCLOS and the redistribution of ocean wealth, *Journal of World Trade Law*, **16** (1982), 27–48.
164. Woodland, A. D. *International Trade and Resource Allocation*. Amsterdam: North-Holland, 1982.

## LIST OF SYMBOLS USED

| | |
|---|---|
| $u^i$ | —utility function |
| $a_i$ | —action |
| $n$ | —number of agents |
| $\mathcal{N}$ | —set of agents |
| $S$ | —subset of N |
| $v$ | —core |
| $A_i$ | —strategy space |
| $\partial$ | —partial derivative |

| | |
|---|---|
| $f$ | —demand function |
| $t$, $\tau$ | —time |
| $\alpha$ | —discount parameter |
| $\sum$ | —summation |
| $\mathcal{N}$ (superscript) | —Nash |
| $V^j$ | —indirect utility function |
| $p$ | —price |
| $m$ | —income |
| $t$ | —tariff |
| $R$ | —reaction locus |
| $r$ | —tariff revenue (Section 4) |
| $e$ | —excess demand |
| $H$ | —indirect trade utility function |
| $\xi$ | —elasticity of excess demand |
| $q$ | —supply |
| $\alpha$, $\beta$, $\gamma$ | —parameters |
| $x$ | —demand |
| $\pi$ | —profit |
| $c$ | —average cost |
| $z$ | —stock |
| $s$ | —resource stock (Section 8) |
| $\eta$ | —demand elasticity |
| $r$ | —extraction rate (Section 8) |
| $\lambda$, $\mu$ | —costate variable |
| $a$, $b$ | —functions |
| $g$ | —gift |
| $G$ | —quantity of public good |
| $\alpha$, $\beta$, $\gamma$, $\theta$, $\zeta$, $\eta$ | —functions (Section 11) |
| $b$ | —benefits (Section 12.1) |
| $s$ | —stake (Section 12.1) |
| $\rho$ | —inflation rate (Section 12.2) |
| $x$ | —excess credit expansion (Section 12.2) |
| $z$, $w$ | —variables (Section 12.2) |
| $\omega$ | —weights |

# INDEX

For Product Safety Concerns and Information please contact our EU
representative GPSR@taylorandfrancis.com
Taylor & Francis Verlag GmbH, Kaufingerstraße 24, 80331 München, Germany

www.ingramcontent.com/pod-product-compliance
Lightning Source LLC
LaVergne TN
LVHW021113300926
849160LV00009B/719
* 9 7 8 0 4 1 5 4 8 8 4 0 2 *